By the same author:

A Furnace (Oxford University Press, 1986)

ROY FISHER

POEMS
1955–1987

Oxford   New York

OXFORD UNIVERSITY PRESS

1988

Oxford University Press, Walton Street, Oxford OX2 6DP
Oxford New York Toronto
Delhi Bombay Calcutta Madras Karachi
Petaling Jaya Singapore Hong Kong Tokyo
Nairobi Dar es Salaam Cape Town
Melbourne Auckland
and associated companies in
Berlin Ibadan

Oxford is a trademark of Oxford University Press

This edition, an expanded edition of Poems 1955–1980 (OUP, 1980),
first published 1988 as an Oxford University Press paperback

British Library Cataloguing in Publication Data
Fisher, Roy, 1930–
Poems 1955–1987.—Expanded ed.—
(Oxford poets).
I. Title
821'.914
ISBN 0–19–282230–6

Library of Congress Cataloging in Publication Data
Fisher, Roy, 1930–
[Poems. Selections]
Poems, 1955–1987 / Roy Fisher.—Expanded ed.
p.   cm.
Previous ed. published as: Poems, 1955–1980. 1980.
I. Title.
821'.914—dc 19   PR6056.I8A6   1988   88–4215
ISBN 0–19–282230–6 (pbk.)

Printed in Great Britain by
J. W. Arrowsmith Ltd., Bristol

For Gael Turnbull

Acknowledgements

The present Oxford Poets paperback edition includes the complete contents of *Poems 1955–1980* (Oxford University Press, 1980), which in turn included the prose work *The Ship's Orchestra* (Fulcrum Press, 1966), *Collected Poems 1968* (Fulcrum Press), *Matrix* (Fulcrum Press, 1971), and *The Thing About Joe Sullivan* (Carcanet New Press, 1978); 'Metamorphoses' and 'Stopped Frames and Set-Pieces' from *The Cut Pages* (Fulcrum Press, 1971); and a group of uncollected poems, some of which had first appeared in *Ikon, Phoenix, Migrant, The Atlantic Review, Aggie Weston's* and Circle Press pamphlets. 'Wonders of Obligation' had first been published by Braad Editions.

Some twenty-four new poems are now added (before *The Ship's Orchestra*), and acknowledgements are due to the editors of the publications in which they severally appeared between 1979 and 1987: *Arc, Bull, Conjunctions, Credences, Diversions (2nd series), Grosseteste Review, Hereford Poems, Lettera, Megaphone, Palantir, Poetry Durham, Poetry Review*, and the *PBS Anthology 1981* (Hutchinson). 'The Elohim Creating Adam' was commissioned for *With a Poet's Eye* (Tate Gallery, 1986), and 'The Whale Knot' for *Whales: a Celebration*. 'The Burning Graves at Netherton' was read on The Poet's Voice (BBC Radio 3).

Contents

POEMS 1955–1987

Aside to a Children's Tale

This dead march is thin
in our spacious street
as the black procession
that stumbles its beat;

our doors are clerical
and the thin coffin door
winks on a wet pall,
a frozen sore;

and four men like pigs
bear high as they can
the unguarded image
of a private man;

while broken music
lamely goes by
in the drummed earth,
the brassy sky.

Why They Stopped Singing

They stopped singing because
They remembered why they had started

Stopped because
They were singing too well

When they stopped they hoped for
A silence to listen into.

Had they sung longer
The people would not have known what to say.

They stopped from the fear
Of singing for ever

They stopped because they saw the rigid world
Become troubled

Saw it begin
Composing a question.

Then they stopped singing
While there was time.

Linear

To travel and feel
the world growing old on your body

breathe and excrete
perpetually the erosion that makes the world

a caravan the little city
that has the wit to cross a continent

so patiently it cannot help but see
how each day's dust lay and shifted and lies again

no forgotten miles or kinks
in the journey other than cunning ones

to pass through many things acquisitively
and touch against many more

a long line without anything
you could call repetition

always through eroded
country amused by others and other worlds

a line like certain snail tracks
crazily long and determined.

Toyland

Today the sunlight is the paint on lead soldiers
Only they are people scattering out of the cool church

And as they go across the gravel and among the spring streets
They spread formality: they know, we know, what they have
                been doing,

The old couples, the widowed, the staunch smilers,
The deprived and the few nubile young lily-ladies,

And we know what they will do when they have opened the doors
                of their houses and walked in:
Mostly they will make water, and wash their calm hands and eat.

The organ's flourishes finish; the verger closes the doors;
The choirboys run home, and the rector goes off in his motor.

Here a policeman stalks, the sun glinting on his helmet-crest;
Then a man pushes a perambulator home; and somebody posts a
                letter.

If I sit here long enough, loving it all, I shall see the District
                Nurse pedal past,
The children going to Sunday School and the strollers strolling;

The lights darting on in different rooms as night comes in;
And I shall see washing hung out, and the postman delivering
                letters.

I might by exception see an ambulance or the fire brigade
Or even, if the chance came round, street musicians (singing and
                playing).

For the people I've seen, this seems the operation of life:
I need the paint of stillness and sunshine to see it that way.

The secret laugh of the world picks them up and shakes them like
                peas boiling;
They behave as if nothing happened; maybe they no longer notice.

I notice. I laugh with the laugh, cultivate it, make much of it,
But I still don't know what the joke is, to tell them.

The Intruder

As I think of yellow,
    a deep shining yellow
that plunges down with white enfolding it,

and as I think of a certain
    stretch of roadway
with a small river beside it in a stone-built channel,

I see, quite clearly, a young girl
    black-haired and quiet
doing some household work, a couple of generations ago.

Her face is broad, like a filbert,
    the features small;
I can see where the colours lie on her skin.

I can see plainly, through the brown varnish
    sentiment lays on the scene
the fat lips and the inexpressive eyes.

A dark body-scent disturbs the air
    but not enough
to persuade me that she and I might sympathize.

She means so little to me, one way or the other
    yet seems so real
it's as if I had walked into somebody else's imagination.

But whose? None of my friends', I think.
    And what will come of it, what
way shall I be involved? The wisest thing

seems to be, by way of the road with the little river
    and the deep shining yellow
to retire discreetly, and leave the sulky bitch to it.

A Debt for Tomorrow

I should sleep now.
    I do not wish it:

even in this pearl silence
the day is not composed to rest as I am.

What is it like, this day
    that I must kill, remorsefully?

A friend. I love him for
his deep simplicity, his understanding
of all I desire or do;
his humour.
        Though hour by hour
steadily his coarseness jostles,
his heavy beauty disappoints and drags
    like a full meal, sleepily;
so now what does he ask for,
of what does he complain? He says
my friendship made a life in him; and now
I'll not stay by to live it out. Surely
if there's a life it's his, his to keep?
How weak it makes him seem, to offer it me!

Necessaries

A spread hand of black trees,
long commands of shade,
old people's polished knuckles
and centuries of timber, thick, lead-veined;

smoke, rolled years of sound
furred into distance; open mouths,
lazy eyes and silver
envelopes of afternoon;

skirted trees and towers
talking like cattle, lengthily.
Gold bells of surrounding days;
stars wheeled into position
while the light lasts, pale.

We need them, you and I; while after dusk
the vans come trundling in, dusting the leaves.

Still our conversation travels
up and down the window panes,
up and down, variously,
while a thin rain film blabs and blears our faces.

The Hospital in Winter

A dark bell leadens the hour,
            the three-o'-clock
light falls amber across a tower.

Below, green-railed within a wall
            of coral brick,
stretches the borough hospital

monstrous with smells that cover death,
            white gauze tongues,
cold-water-pipes of pain, glass breath,

porcelain, blood, black rubber tyres;
            and in the yards
plane trees and slant telephone wires.

On benches squat the afraid and cold
            hour after hour.
Chains of windows snarl with gold.

Far off, beyond the engine-sheds,
            motionless trucks
grow ponderous, their rotting reds

deepening towards night; from windows
            bathrobed men
watch the horizon flare as the light goes.

Smoke whispers across the town,
            high panes are bleak;
pink of coral sinks to brown;
a dark bell brings the dark down.

Leaving July

Low crippled clouds drag on a naked sky
over night leaves that point
ravines of darkest green down steeply
from the pale plateau of glaucous twilight;

the sky flattens on the land and gazes
back up into itself with rainwater eyes
out of blue rutted sockets on a builder's site:

it levels along the wires and the stump arms
of telegraph poles, almost at cool-tiled house-height;
where long roofs make a floor for shallow midnight.

Five Morning Poems from a Picture by Manet

1
Wells of shadow by the stream
and the long branches
high over pools of gold light in the grass.

A drift of morning scents
downward through pines,
dazzled, to hang on water finally;
the road's deserted curve; beyond,
blue town-smoke standing.

Schoolboy in the scarlet cap,
leaning across the wall
motionless —
will you feel the moment end,
a dead scale sliding from a fish?

Plain-featured boy, cast in a pose of beauty.

Wells of shadow spread the stream
so dark it glides invisibly
and leaves his coat-cuffs dry.

Slant eyes that shine too deep a brown
to show time, or to tell
what tautens
stillness round him to a fine
film that shivers, glazed and sombre:

the scents, maybe, that sting.

or ironstone in the water, the taste of April.

So dark it glides invisibly
the moment takes its form
out of his smile, that rests on turbulence —
there have been tears, there could be greed.

Someone has given him cherries:
untouched upon the wall
they spill from yellow paper, crimson-heavy;
glossy, their soft skins burst.

High over light-pools in the grass
and still boughs, a bronze shadow
gestures behind my dream,
a god, polished to nothing by our thoughts:

plenitude, rising at the heart's pace to noon,
that shadow turns clear waterglass of light
to amber, and extinguishes
the last white flecks of agitation.

Music of the generous eye,
swelling, receding on my breath.

Here are the golden distances;
deeper than life the stream goes in the shade.

Cool trees. Clarity. Dreaming boy.
Fat cherries and a scarlet cap;
water under the thin smoke of morning —

music of the generous eye.

2

Nowhere. Hear the stormy spring
far off across the racks of quiet. Nothing.
Suddenly, a leaning fence-post daubed with rain.

It rots into the day. Cold lights
swell from the clouds, then take their death away.

Some ceremony lets fall its links:
a glistening fence-post suddenly encountered;

iron trumpets in the sky.

All in the flooded meadow, ankle deep,
the lately dead, the tidy men and women
walking with steps like flakes of agony.

Accurate as the trumpet-notes they go,
with polished shoes in the gunmetal water;
they tread on loss as once they counted money.

All in the meadow. Flakes of agony
walking with marble steps to iron trumpets.

Ceremony to wear away the world.

Barbaric news for someone about to die:
each footfall muffles breath, each iron grace-note
stoppers the heart with heartless beauty.

A world freezing apart, where a movement
cracks the glass veins of music in the flesh.

Islands of understanding float, confused:
rain-smears on rotting wood, the sheepish dead
who stand amid the floods, bright stems of glass
shattering by inches; lakes of misery —

The iron trumpets wear away the sky.

3

Death music sounds for a boy by a stream,
urgent as waves, as polished shoes,
though amber screens of light shade him from time.

Ironstone in the water, tasting
coldly of April pleasure, yet
embitters like brass the pools of golden morning.

Trumpets of iron shake the sky;
in chains of ceremony they spill
out of the music of a generous eye.

Over his quiet they wear away
all to a frigid emphasis
on lines that form his child's flesh into beauty.

Slowly comes the polished step,
cut marble agony to blight
the charm of cherries and the scarlet cap;

to drag the forms of beauty till they gleam
dully on lakes of misery
rising from wells of shadow by the stream.

4

Looking for life, I lost my mind:
only the dead
spoke through their yellow teeth
into the marble tombs they lay beneath,
splinters of fact stuck in the earth's fat rind.

The privet glittered with noon; the hill
slid its green head
down from the sky to try me. I was still
as the evil-shaped dark leaves.
Then I heard what the corpses said.

Muttering, they told me how their lives
from burial
spiked back at the world like knives
striking the past for legacies of wrong —
the fiction of understanding worst of all.

I saw a vase of familiar flowers,
behind it, a tray
of thin pale brass with patterned borders;
the troubling taste of an alien mouth not figured
even in fantasy.

5
Walking beside the wall towards the stream
where flickering water-lights
lick amber tongues quickly across stone,
with a yellow bag of cherries, an old jacket,
a red cap crumpled on thick hair,
thought, an uneven slime;
gravel underfoot, the wall's rough edge;
stub fingers sliding over pads of moss.

Stout morning walls the sun walks down,
smoke bursting slow into invisibility,
the tight sky broken among banks of trees;

sounds vanishing through worn patches on the road,
dark silence dropping from the ledges of the pines.

Flat heads of stone amid the turf. The shadowy stream.

To the Memory of Wyndham Lewis

A trellis of clean men and painted steel
straggles above the tide lines. Autumn day;
an eastern shingle beach
where, caliper-wise, sour schemes confront the surf.

Under the clouds' rebellious plain
drab uniforms peg out their barricades;
like the squat lorries, noon hours pass.

As moorland bushes take the wind
thoughts lean all day pale-eyed towards the sea;
spikes of dune grass stand lithe,
flickering beneath the searchlights' stone-blind gaze.

In the raw, bloodless rampart,
humanity, a turbid stream,
runs dark amid the diagrams and the helmets,

while along oiled steel grooves,
in cruel harness with the uneasy hours,
systems advance on night.

2

The white mare rolls on the pebbles at the edge of the sea.
Searchlights banked behind helmet rims
burn roads across the shelving beach.
The rolling-mill of the night
endlessly sucks the dark in from the land
with a finger of foam to its disappearing lips,
while the white mare on the brilliant pebble shore rolls,
helmet-bowls converse, and in
the red shadow of the dune-grass bayonets
glint on the thought of blood petals.

The horse's stony teeth are young and perishable
as ripples vanishing searchlit among rocks;
the soldiery smoke-stain her coat: to be
near them yellows white; their presence taints
the metal flood rising beyond the surf.

O sepia-blooded soldiers with tobacco coats,
you have waited long; your armoured heads
no more seem to be moving by the bright beams.
The middle darkness is seething round you fast
into the foam; and the mare will not be still
down on those stones you have made glisten so fiercely.

3

The captain to the white mare. 'Handsome thing,
our crescent net entraps you; waves that freeze
thunder along the shore between its horns.
Soldiers tire, hours sag;
the toothed wheel of a stopwatch wakes the rifles.
You will be torn and shattered; we shall rake
with bayonets in your body.
What's left of you will be, by morning,
shovelled into the sea. We shall remember you.

12

'I think your blood will rot the world from under us,
melting into an ordure of decay —

'Enough. That I'll survive. I see
rising above that filth a scaffolding:
Expediency. There at the last I'll wither.'

4
Her eyes like staring lamps of stone,
she charged the ramp, cantered on flying pebbles
to rear magnesium-white in searchlight rays.

Everyone scattered. The lights blinded the mare,
till slowly, unguided on aimless silent bearings,
they flared their square weight down along empty strand.

Then she was gone, thudding in fronds of dark,
beyond the buttoned hearts of soldiers
who stood all ways, dull basin-hats
tilted askew, their round eyes frozen beads.

Flung into stupefaction by her flight
their taut-strapped faces gaped
in the long concrete jaw of the sea wall;
great bars of yellow light still striped the beach,
the sea teasing and wrenching at their tips.

Voices. Bayonets clattering in the dunes;
brown greatcoats looming; light-beams
snapped into nothing. Vacant order forming.

CITY

City

On one of the steep slopes that rise towards the centre of the city all the buildings have been destroyed within the past year: a whole district of the tall narrow houses that spilled around what were a hundred years ago outlying factories has gone. The streets remain, among the rough quadrilaterals of brick rubble, veering awkwardly towards one another through nothing; at night their rounded surfaces still shine under the irregularly-set gaslamps, and tonight they dully reflect also the yellowish flare, diffused and baleful, that hangs flat in the clouds a few hundred feet above the city's invisible heart. Occasional cars move cautiously across this waste, as if suspicious of the emptiness; there is little to separate the roadways from what lies between them. Their tail-lights vanish slowly into the blocks of surrounding buildings, maybe a quarter of a mile from the middle of the desolation.

And what is it that lies between these purposeless streets? There is not a whole brick, a foundation to stumble across, a drainpipe, a smashed fowlhouse; the entire place has been razed flat, dug over, and smoothed down again. The bald curve of the hillside shows quite clearly here, near its crown, where the brilliant road, stacked close on either side with warehouses and shops, runs out towards the west. Down below, the district that fills the hollow is impenetrably black. The streets there are so close and so twisted among their massive tenements that it is impossible to trace the line of a single one of them by its lights. The lamps that can be seen shine oddly, and at mysterious distances, as if they were in a marsh. Only the great flat-roofed factory shows clear by its bulk, stretching across three or four whole blocks just below the edge of the waste, with solid rows of lit windows.

Lullaby and Exhortation for the Unwilling Hero

A fish,
Firelight,
A watery ceiling:
Under the door
The drunk wind sleeps.

The bell in the river,
The loaf half eaten,
The coat of the sky.

A pear,
Perfume,
A white glade of curtains:
Out in the moonlight
The smoke reaches high.

The statue in the cellar,
The skirt on the chairback,
The throat of the street.

A shell,
Shadow,
A floor spread with silence:
Faint on the skylight
The fat moths beat.

The pearl in the stocking,
The coals left to die,
The bell in the river,
The loaf half eaten,
The coat of the sky.

The night slides like a thaw
And oil-drums bang together.

A frosted-glass door opening, then another.
Orange and blue *décor*.
The smoke that hugs the ceiling tastes of pepper.

What steps descend, what rails conduct?
Sodium bulbs equivocate,
And cowls of ventilators
With limewashed breath hint at the places
To which the void lift cages plunge or soar.

Prints on the landing walls
Are all gone blind with steam;
A voice under the floor
Swings a dull axe against a door.

The gaping office block of night
Shudders into the deep sky overhead:

Thrust down your foot in sleep
Among its depths. Do not respect
The janitors in bed,
The balustrades of iron bars,
The gusty stairwells; thrust it deep,
Into a concrete garage out of sight,
And rest among the cars
That, shut in filtered moonlight,
Sweat mercury and lead.

Subway trains, or winds of indigo,
Bang oil-drums in the yard for war:
Already, half-built towers
Over the bombed city
Show mouths that soon will speak no more,
Stoppered with the perfections of tomorrow.

You can lie women in your bed
With glass and mortar in their hair.
Pocket the key, and draw the curtains,
They'll not care.

Letters on a sweetshop window:
At last the rain slides them askew,
The foetus in the dustbin moves one claw.

And from the locomotive
That's halted on the viaduct
A last white rag of steam
Blows ghostly across the gardens.
When you wake, what will you do?

Under the floorboards of your dream
Gun barrels rolled in lint
Jockey the rooms this way and that.
Across the suburbs, squares of colour gleam:
Swaddled in pink and apricot,
The people are 'making love'.

Those are bright points that flicker
Softly, and vanish one by one.

Your telegraphic fingers mutter the world.
What will they reach for when your sleep is done?

The hiss of tyres along the gutter,
Odours of polish in the air;
A car sleeps in the neighbouring room,
A wardrobe by its radiator.

The rumbling canisters beat for you
Who are a room now altogether bare,
An open mouth pressed outwards against life,
Tasting the sleepers' breath,
The palms of hands, discarded shoes,
Lilac wood, the blade of a breadknife.

Before dawn in the sidings,
Over whose even tracks
Fat cooling towers caress the sky,
The rows of trucks
Extend: black, white,
White, grey, white, black,
Black, white, black, grey,
Marshalled like building blocks:

Women are never far away.

In the century that has passed since this city has become great, it has twice laid itself out in the shape of a wheel. The ghost of the older one still lies among the spokes of the new, those dozen highways that thread constricted ways through the inner suburbs, then thrust out, twice as wide, across the housing estates and into the countryside, dragging moraines of buildings with them. Sixty or seventy years ago there were other main roads, quite as important as these were then, but lying between their paths. By day they are simply alternatives, short cuts, lined solidly with parked cars and crammed with delivery vans. They look merely like side-streets, heartlessly overblown in some excess of Victorian expansion. By night, or on a Sunday, you can see them for what they are. They are still lit meagrely, and the long rows of houses, three and four storeys high, rear black above the lamps enclosing the road-ways, clamping them off from whatever surrounds them. From these pavements you can sometimes see the sky at night, not obscured as it is in most parts of the city by the greenish-blue haze of light that steams out of the mercury vapour lamps. These streets are not worth lighting. The houses have not been turned into shops — they are not villas either that might have become offices, but simply tall dwellings, opening straight off the street, with cavernous entries leading into back courts.

The people who live in them are mostly very old. Some have lived through three wars, some through only one; wars of newspapers, of mysterious sciences, of coercion, of disappearance. Wars that have come down the streets from the unknown city and the unknown world, like rainwater floods in the gutters. There are small shops at street corners, with blank rows of houses between them; and taverns carved only shallowly into the massive walls. When these people go into the town, the buses they travel in stop just before they reach it, in the sombre back streets behind the Town Hall and the great insurance offices.

These lost streets are decaying only very slowly. The impacted lives of their inhabitants, the meaninglessness of news, the dead black of the chimney breasts, the conviction that the wind itself comes only from the next street, all wedge together to keep destruction out; to deflect the eye of the developer. And when destruction comes, it is total: the printed notices on the walls, block by block, a few doors left open at night, broken windows advancing down a street until fallen slates appear on the pavement and are not kicked away. Then, after a few weeks of this, the machines arrive.

The Entertainment of War

I saw the garden where my aunt had died
And her two children and a woman from next door;
It was like a burst pod filled with clay.

A mile away in the night I had heard the bombs
Sing and then burst themselves between cramped houses
With bright soft flashes and sounds like banging doors;

The last of them crushed the four bodies into the ground,
Scattered the shelter, and blasted my uncle's corpse
Over the housetop and into the street beyond.

Now the garden lay stripped and stale; the iron shelter
Spread out its separate petals around a smooth clay saucer,
Small, and so tidy it seemed nobody had ever been there.

When I saw it, the house was blown clean by blast and care:
Relations had already torn out the new fireplaces;
My cousin's pencils lasted me several years.

And in his office notepad that was given me
I found solemn drawings in crayon of blondes without dresses.
In his lifetime I had not known him well.

Those were the things I noticed at ten years of age:
Those, and the four hearses outside our house,
The chocolate cakes, and my classmates' half-shocked envy.

But my grandfather went home from the mortuary
And for five years tried to share the noises in his skull,
Then he walked out and lay under a furze-bush to die.

When my father came back from identifying the daughter
He asked us to remind him of her mouth.
We tried. He said 'I think it was the one'.

These were marginal people I had met only rarely
And the end of the whole household meant that no grief was seen;
Never have people seemed so absent from their own deaths.

This bloody episode of four whom I could understand better dead
Gave me something I needed to keep a long story moving;
I had no pain of it; can find no scar even now.

But had my belief in the fiction not been thus buoyed up
I might, in the sigh and strike of the next night's bombs
Have realized a little what they meant, and for the first time been afraid.

North Area

Those whom I love avoid all mention of it,
Though certain gestures they've in common
Persuade me they know it well:
A place where I can never go.

No point in asking why, or why not.
I picture it, though —
There must be dunes with cement walks,
A twilight of aluminium
Among beach huts and weather-stained handrails;
Much glass to reflect the clouds;
And a glint of blood in the cat-ice that holds the rushes.

The edge of the city. A low hill with houses on one side and rough common land on the other, stretching down to where a dye-works lies along the valley road. Pithead gears thrust out above the hawthorn bushes; everywhere prefabricated workshops jut into the fields and the allotments. The society of singing birds and the society of mechanical hammers inhabit the world together, slightly ruffled and confined by each other's presence.

By the Pond

This is bitter enough: the pallid water
With yellow rushes crowding toward the shore,
That fishermen's shack.

The pit-mound's taut and staring wire fences,
The ashen sky. All these can serve as conscience.
For the rest, I'll live.

Brick-dust in sunlight. That is what I see now in the city, a dry epic flavour, whose air is human breath. A place of walls made straight with plumbline and trowel, to dessicate and crumble in the sun and smoke. Blistered paint on cisterns and girders, cracking to show the priming. Old men spit on the paving slabs, little boys urinate; and the sun dries it as it dries out patches of damp on plaster facings to leave misshapen stains. I look for things here that make old men and dead men seem young. Things which have escaped, the landscapes of many childhoods.

Wharves, the oldest parts of factories, tarred gable ends rearing to take the sun over lower roofs. Soot, sunlight, brick-dust; and the breath that tastes of them.

At the time when the great streets were thrust out along the old high-roads and trackways, the houses shouldering towards the country and the back streets filling in the widening spaces between them like webbed membranes, the power of will in the town was more open, less speciously democratic, than it is now. There were, of course, cottage railway stations, a jail that pretended to be a castle out of Grimm, public urinals surrounded by screens of cast-iron lacework painted green and scarlet; but there was also an arrogant ponderous architecture that dwarfed and terrified the people by its sheer size and functional brutality:

the workhouses and the older hospitals, the thick-walled abattoir, the long vaulted market-halls, the striding canal bridges and railway viaducts. Brunel was welcome here. Compared with these structures the straight white blocks and concrete roadways of today are a fairground, a clear dream just before waking, the creation of salesmen rather than of engineers. The new city is bred out of a hard will, but as it appears, it shows itself a little ingratiating, a place of arcades, passages, easy ascents, good light. The eyes twinkle, beseech and veil themselves; the full, hard mouth, the broad jaw — these are no longer made visible to all.

A street half a mile long with no buildings, only a continuous embankment of sickly grass along one side, with railway signals on it, and strings of trucks through whose black-spoked wheels you can see the sky; and for the whole length of the other a curving wall of bluish brick, caked with soot and thirty feet high. In it, a few wicket gates painted ochre, and fingermarked, but never open. Cobbles in the roadway.

A hundred years ago this was almost the edge of town. The goods yards, the gasworks and the coal stores were established on tips and hillocks in the sparse fields that lay among the houses. Between this place and the centre, a mile or two up the hill, lay a continuous huddle of low streets and courts, filling the marshy valley of the meagre river that now flows under brick and tarmac. And this was as far as the railway came, at first. A great station was built, towering and stony. The sky above it was southerly. The stately approach, the long curves of wall, still remain, but the place is a goods depot with most of its doors barred and pots of geraniums at those windows that are not shuttered. You come upon it suddenly in its open prospect out of tangled streets of small factories. It draws light to itself, especially at sunset, standing still and smooth faced, looking westwards at the hill. I am not able to imagine the activity that must once have been here. I can see no ghosts of men and women, only the gigantic ghost of stone. They are too frightened of it to pull it down.

The Sun Hacks

The sun hacks at the slaughterhouse campanile,
And by the butchers' cars, packed tail-to-kerb,
Masks under white caps wake into human faces.

The river shudders as dawn drums on its culvert;
On the first bus nightworkers sleep, or stare
At hoardings that look out on yesterday.

The whale-back hill assumes its concrete city:
The white-flanked towers, the stillborn monuments;
The thousand golden offices, untenanted.

At night on the station platform, near a pile of baskets, a couple embraced, pressed close together and swaying a little. It was hard to see where the girl's feet and legs were. The suspicion this aroused soon caused her hands, apparently joined behind her lover's back, to become a small brown paper parcel under the arm of a stout engine-driver who leaned, probably drunk, against the baskets, his cap so far forward as almost to conceal his face. I could not banish the thought that what I had first seen was in fact his own androgynous fantasy, the self-sufficient core of his stupor. Such a romantic thing, so tender, for him to contain. He looked more comic and complaisant than the couple had done, and more likely to fall heavily to the floor.

A café with a frosted glass door through which much light is diffused. A tall young girl comes out and stands in front of it, her face and figure quite obscured by this milky radiance.

She treads out on to a lopsided ochre panel of lit pavement before the doorway and becomes visible as a coloured shape, moving sharply. A wrap of honey and ginger, a flared saffron skirt, grey-white shoes. She goes off past the Masonic Temple with a young man: he is pale, with dark hair and a shrunken, earnest face. You could imagine him a size larger. Just for a moment, as it happens, there is no one else in the street at all. Their significance escapes rapidly like a scent, before the footsteps vanish among the car engines.

A man in the police court. He looked dapper and poker-faced, his arms straight, the long fingers just touching the hem of his checked jacket. Four days after being released from the prison where he had served two years for theft he had been discovered at midnight clinging like a tree-shrew to the bars of a glass factory-roof. He made no attempt to explain his presence there; the luminous nerves that made him fly up to it were not visible in daylight, and the police seemed hardly able to believe this was the creature they had brought down in the darkness.

In this city the governing authority is limited and mean: so limited that it can do no more than preserve a superficial order. It supplies fuel, water and power. It removes a fair proportion of the refuse, cleans the streets after a fashion, and discourages fighting. With these things, and a few more of the same sort, it is content. This could never be a capital city for all its size. There is no mind in it, no regard. The sensitive, the tasteful, the fashionable, the intolerant and powerful, have not moved through it as they have moved through London, evaluating it, altering it deliberately, setting in motion wars of feeling about it. Most of it has never been seen.

In an afternoon of dazzling sunlight in the thronged streets, I saw at first no individuals but a composite monster, its unfeeling surfaces matted with dust: a mass of necks, limbs without extremities, trunks without heads; unformed stirrings and shovings spilling across the streets it had managed to get itself provided with.

Later, as the air cooled, flowing loosely about the buildings that stood starkly among the declining rays, the creature began to divide and multiply. At crossings I could see people made of straws, rags, cartons, the stuffing of burst cushions, kitchen refuse. Outside the Grand Hotel, a long-boned carrot-haired girl with glasses, loping along, and with strips of bright colour, rich, silky green and blue, in her soft clothes. For a person made of such scraps she was beautiful.

Faint blue light dropping down through the sparse leaves of the plane trees in the churchyard opposite after sundown, cooling and shaping heads, awakening eyes.

The Hill behind the Town

Sullen hot noon, a loop of wire,
With zinc light standing everywhere,
A glint on the chapels,
Glint on the chapels.

Under my heel a loop of wire
Dragged in the dust is earth's wide eye,
Unseen for days,
Unseen days.

Geranium-wattled, fenced in wire,
Caged white cockerels crowd near
And stretch red throats,
Stretch red throats;

Their cries tear grievous through taut wire,
Drowned in tanks of factory sirens
At sullen noon,
Sullen hot noon.

The day's on end; a loop of wire
Kicked from the dust's bleak daylight leaves
A blind white world,
Blind white world.

The Poplars

Where the road divides
Just out of town
By the wall beyond the filling-station
Four Lombardy poplars
Brush stiff against the moorland wind.

Clarity is in their tops
That no one can touch
Till they are felled,
Brushwood to cart away:

To know these tall pointers
I need to withdraw
From what is called my life
And from my net
Of achievable desires.

Why should their rude and permanent virginity
So capture me? Why should studying
These lacunae of possibility
Relax the iron templates of obligation
Leaving me simply Man?

All I have done, or can do
Is prisoned in its act:
I think I am afraid of becoming
A cemetery of performance.

Starting to Make a Tree

First we carried out the faggot of steel stakes; they varied in length, though most were taller than a man.

We slid one free of the bundle and drove it into the ground, first padding the top with rag, that the branch might not be injured with leaning on it.

Then we took turns to choose stakes of the length we wanted, and to feel for the distances between them. We gathered to thrust them firmly in.

There were twenty or thirty of them in all; and when they were in place we had, round the clearing we had left for the trunk, an irregular radial plantation of these props, each with its wad of white at the tip. It was to be an old, downcurving tree.

This was in keeping with the burnt, chemical blue of the soil, and the even hue of the sky which seemed to have been washed with a pale brownish smoke;

another clue was the flatness of the horizon on all sides except the north, where it was broken by the low slate or tarred shingle roofs of the houses, which stretched away from us for a mile or more.

This was the work of the morning. It was done with care, for we had no wish to make revisions;

we were, nonetheless, a little excited, and hindered the women at their cooking in our anxiety to know whose armpit and whose groin would help us most in the modelling of the bole, and the thrust of the boughs.

That done, we spent the early dusk of the afternoon gathering materials from the nearest houses; and there was plenty:

a great flock mattress; two carved chairs; cement; chicken-wire; tarpaulin; a smashed barrel; lead piping; leather of all kinds; and many small things.

In the evening we sat late, and discussed how we could best use them. Our tree was to be very beautiful.

Yet whenever I see that some of these people around me are bodily in love, I feel it is my own energy, my own hope, tension and sense of time in hand, that have gathered and vanished down that dark drain; it is I who am left, shivering and exhausted, to try and kick the lid back into place so that I can go on without the fear of being able to feel only vertically, like a blind wall, or thickly, like the tyres of a bus.

Lovers turn to me faces of innocence where I would expect wariness. They have disappeared for entire hours into the lit holes of life, instead of lying stunned on its surface as I, and so many, do for so long; or instead of raising their heads cautiously and scenting the manifold airs that blow through the streets.

The city asleep. In it there are shadows that are sulphurous, tanks of black bile. The glitter on the roadways is the deceptive ore that shines on coal.

The last buses have left the centre; the pallid faces of the crowd looked like pods, filled by a gusty summer that had come too late for plenty.

Silvered rails that guide pedestrians at street corners stand useless. Towards midnight, or at whatever hour the sky descends with its full iron weight, the ceilings drop lower everywhere; each light is partial, and proper only to its place. There is no longer any general light, only particular lights that overlap.

Out of the swarming thoroughfares, the night makes its own streets with a rake that drags persuaded people out of its way: streets where the bigger buildings have already swung themselves round to odd angles against the weakened currents of the traffic.

There are lamplit streets where the full darkness is only in the deep drains and in the closed eyesockets and shut throats of the old as they lie asleep; their breath moves red tunnel-lights.

The main roads hold their white-green lights with difficulty, like long, loaded boughs; when the machines stop moving down them their gradients reappear.

Journeys at night: sometimes grooves in a thick substance, sometimes raised weals on black skin.

The city at night has no eye, any more than it has by day, although you would expect to find one; and over much of it the sleep is aqueous and incomplete, like that of a hospital ward.

But to some extent it stops, drops and congeals. It could be broken like asphalt, and the men and women rolled out like sleeping maggots.

Once I wanted to prove the world was sick. Now I want to prove it healthy. The detection of sickness means that death has established itself as an element of the timetable; it has come within the range of the measurable. Where there is no time there is no sickness.

The Wind at Night

The suburb lies like a hand tonight,
A man's thick hand, so stubborn
No child or poet can move it.

The wind drives itself mad with messages,
Clattering train wheels over the roofs,
Collapsing streets of sound until
Far towers, daubed with swollen light,
Lunge closer to abuse it,

This suburb like a sleeping hand,
With helpless elms that shudder
Angry between its fingers,
Powerless to disprove it.

And, although the wind derides
The spaces of this stupid quarter,
And sets the time of night on edge,
It mocks the hand, but cannot lose it:

This stillness keeps us in the flesh,
For us to use it.

I stare into the dark; and see a window, a large sash window of four panes, such as might be found in the living-room of any fair-sized old house. Its curtains are drawn back and it looks out on to a small damp garden, narrow close at hand where the kitchen and outhouses lead back, and then almost square. Privet and box surround it, and the flowerbeds are empty save for a few laurels or rhododendrons, some leafless rose shrubs and a giant yucca. It is a December afternoon, and it is raining. Not far from the window is a black marble statue of a long-haired, long-bearded old man. His robes are conventionally archaic, and he sits, easily enough, on what seems a pile of small boulders, staring intently and with a look of great intelligence towards the patch of wall just under the kitchen window. The statue looks grimy, but its exposed surfaces are highly polished by the rain, so that the nose and the cheek-bones stand out strongly in the gloom. It is rather smaller than life-size. It is clearly not in its proper place: resting as it does across the moss of the raised border, it is appreciably tilted forward and to one side, almost as if it had been abandoned as too heavy by those who were trying to move it — either in or out.

Walking through the suburb at night, as I pass the dentist's house I hear a clock chime a quarter, a desolate brassy sound. I know where it stands, on the mantelpiece in the still surgery. The chime falls back into the house, and beyond it, without end. Peace.

I sense the simple nakedness of these tiers of sleeping men and women beneath whose windows I pass. I imagine it in its own setting, a mean bathroom in a house no longer new, a bathroom with plank panelling, painted a peculiar shade of green by an amateur, and badly preserved. It is full of steam, so much as to obscure the yellow light and hide the high, patched ceiling. In this dream, standing quiet, the private image of the householder or his wife, damp and clean.

I see this as it might be floating in the dark, as if the twinkling point of a distant street-lamp had blown in closer, swelling and softening to a foggy oval. I can call up a series of such glimpses that need have no end, for they are all the bodies of strangers. Some are deformed or diseased, some are ashamed, but the peace of humility and weakness is there in them all.

I have often felt myself to be vicious, in living so much by the eye, yet among so many people. I can be afraid that the egg of light through which I see these bodies might present itself as a keyhole. Yet I can find no sadism in the way I see them now. They are warm-fleshed, yet their shapes have the minuscule, remote morality of some mediaeval woodcut of the Expulsion: an eternally startled Adam, a permanently bemused Eve. I see them as homunculi, moving privately each in a softly lit fruit in a nocturnal tree. I can consider without scorn or envy the well-found bedrooms I pass, walnut and rose-pink, altars of tidy, dark-haired women, bare-backed, wifely. Even in these I can see order.

I come quite often now upon a sort of ecstasy, a rag of light blowing among the things I know, making me feel I am not the one for whom it was intended, that I have inadvertently been looking through another's eyes and have seen what I cannot receive.

I want to believe I live in a single world. That is why I am keeping my eyes at home while I can. The light keeps on separating the world like a table knife: it sweeps across what I see and suggests what I do not. The imaginary comes to me with as much force as the real, the remembered with as much force as the immediate. The countries on the map divide and pile up like ice-floes: what is strange is that I feel no stress, no grating discomfort among the confusion, no loss; only a belief that I should not be here. I see the iron fences and the shallow ditches of the countryside the mild wind has travelled over. I cannot enter that countryside; nor can I escape it. I cannot join together the mild wind and the shallow ditches, I cannot lay the light across the world and then watch it slide away. Each thought is at once translucent and icily capricious. A polytheism without gods.

The Park

If you should go there on such a day —
The red sun disappearing,
Netted behind black sycamores;

If you should go there on such a day —
The sky drawn thin with frost,
Its cloud-rims bright and bitter —

If you should go there on such a day,
Maybe the old goose will chase you away.

If you should go there to see
The shallow concrete lake,
Scummed over, fouled with paper;

If you should go there to see
The grass plots, featureless,
Muddy, and bruised, and balding —

If you should go there to see,
Maybe the old goose will scare you as he scared me

Waddling fast on his diseased feet,
His orange bill thrust out,
His eyes indignant;

Waddling fast on his diseased feet,
His once-ornamental feathers
Baggy, and smeared with winter —

Waddling fast on his diseased feet,
The old goose will one day reach death; and be unfit to eat.

And when the goose is dead, then we
Can say we're able, at last,
No longer hindered from going;

And when the goose is dead, then we
Have the chance, if we still want it,
To wander the park at leisure;

— Oh, when excuse is dead, then we
Must visit there, most diligently.

Chirico

'What I listen to is worthless. There is no mystery in music'
Giorgio de Chirico

Let the flute
be silenced
and the breath
drawn back and photographed
a word.

And the stone
wagons
at rest on my ears
to quieten. My ears know
how to read.

My eyes
know how to
count each other
sleep or wake. Let the flute
be entered.

Be spoken
be used
imprinted
in a man's spine or a roof
or cloud.

Let the flute
be no flute
and the mouth
drawn back to talk
about teeth.

There are stairs
in cylinder
and for notes
posters of insurrection
print ebony.

And the clock
contends
even with masonry. Through
stone I hear it emptying
like a washbowl.

When time
is a hole
the musician loses
his notes through cracks between
the paving stones.

Let each one
be deaf
and the tone
damped under the slab. I read
a typewriter of stone.

STOPPED FRAMES AND SET-PIECES

As he heaved himself backwards into the water, arms above his head, and went under, for an instant the refraction in the broken water shattered the shapes of his belly, torso and bearded face, blurring and magnifying the whole so that it looked like a crucified Christ coming apart like a cloud figure. That was for half a second: then as the shape, sinking and waving, showed a shade more tranquil, it was more like God Himself, uncomfortably waking.

The god Mercury, plump with winter, flew steadily through the black, gritty dawn, among the pigeon-spattered cornices of one of the cities of Germany.

Furious, the boxer dog leapt in pursuit and was brought up short by its lead while in the air, only one foot touching the ground. The strength of its impetus, the whole weight of the body yanking on the strap, distorted it: the head seemed as if it was being twisted off, and the creature for all its heaviness and force had no more stability of shape than a slug hurled at a paving slab.

In the second the dog's form was destroyed and before it reconstituted itself a fallen scrambling dog, it had an alternative form of paroxysm, a sudden precipitate out of energy, a bigger payout than was budgeted; dog-sized, big enough to remember, bursting out from the subsiding animal, twisting on into the world, detaching itself, a single plasm, weightless and self-determining.

A small broken Indian statue, of about the thirteenth century. The figure of a girl dancer, or, rather, the torso, for the head and neck are missing, both arms are cracked off just below the shoulders, and the legs are both lost, the left finishing just above the knee, the other a little higher. The right breast has been smashed, the nipple of the other worn away.

Even without the extremities that will have shown what it was intended to mean, the body, with its long supple waist and round thighs, is graceful. The shoulders are held a little way back, and the belly is thrust forward, swung to one side. There's no temptation to guess at the positions of the vanished head and limbs.

It is clothed only in festoons of decorations, metal and stones against the soft flesh which now, its stone polish transformed by age into myriad roughenings and craters, looks porous and sleepy. There is a broad collar with heavy pendants, one of which, a triple rope of what seem to be pearls, falls sidelong to the waist. A necklace of similar design is cast loosely about the shoulders and falls through the deep hollow of the bosom. Just above the break in the left upper arm there is a tight circlet that rises in the shape of a crown. The hips are circled by a low deep belt that reaches to the loins: it starts with a tight flat band, of gold maybe, chased with circling patterns and then loosens into thick swathes and bands, of intricate design, falling low at the front and gathered up neatly at the hips. From this belt hang a central tasselled loop, heavy with baubles, that passes freely between the legs, and also the decorations of the thighs, pendants of beaten metal stretching down toward the vanished knees. Dangling between these and the belt are heavy braided loops, hung with medallions, several bands of them, the higher ones loosely curved, the lower clasping the limbs more closely.

Without gestures or face, this body is dominated, not by its single breast, but by the large and beautiful navel, thrust out and held up into a pout by the constricting belt that curves close beneath it. It is very slightly oval, stretched not upwards, but out across the belly. If the figure were life-size, the palm of a hand could only just cover its hollow, that begins very gradually, then plunges evenly into a deep shadow, ridged at the upper edge by a last flat fold of skin. This is the only feature the body possesses. It is eye, or mouth, or anus, or ear: all the body's orifices. In itself it is useless, a scar crater, an insignia of coming into life; a gross and handsome decoration riding among the goldsmiths' ornaments.

The spray of bullets had struck across the people walking on the pavement. Most rushed for shelter and a few lay still. A big man in a tight suit, lying in the gutter with no apparent injury, began to sit up, clearly conscious of his terribly exposed situation.

The big man in the gutter was up on his feet and tiptoeing across the sunlit pavement. The gunmen, if they were still there, let him go, treading in other people's blood, to get to a shop doorway.

Encouraged by the big man's example, men and women were picking themselves slowly up and stumbling for cover without looking round. From the dead there were long runs of blood, right down into the gutters. Its brightness was astonishing, the gaiety of the colour.

A bare isthmus of ash projecting into the bay, and ending in a little mountain of the most ominous shape, quite shallow, a flat-topped cone, almost a simple geometrical form except for the slight tilt to one side of the platform at the top. The sides sloped evenly down into the motionless and slaty sea. There was no vegetation anywhere about. The sun was just below the long black perfectly horizontal cliffs across the bay.

Along a little causeway a strange domed house had been built on piles in the water. It had balconies of cast iron lacework and big windows, most of them now broken.

The Gentleman's Patent Reducing Bath was a polished cabinet lined with zinc and mounted on legs like a stool. On one side of the cabinet there was an opening, shaped rather like a proscenium, from which hung not a curtain but the legs of a kind of rubber bloomers, bonded to the zinc inside the cabinet. The gentleman, stripping himself naked, would step from a chair into the cabinet, easing himself down until his legs, thighs encased in the bloomers which gripped firmly above the knee, projected through the opening and he was able to rest his feet on the ground. His servant would then pour jugs of warm water into the cabinet. The legs of the gentleman's drawers would fill first; and so on, until he was immersed from knee to waist. A stout wooden cover, the two halves of which fitted together to leave a hole like that of a lavatory seat only larger, completed the apparatus, giving him a dry rest for his elbows and preventing him from dropping his book into the water.

As he stood there in the sunlight, his body placid, the judge's robes and wig were unruffled, by even the least breath, and his heavy face was at rest. Yet this face was set as if in a gale that tore across it from right to left, pulling at the flesh, distorting the eyes and mouth.

The snow turned into the mist; the horizon was so close they couldn't see it. The other four came out of the mist, grey and bearlike. Edmonds was dancing along with the great orange balloon over his head, pretending it was carrying him along; Ray, recognizable by his movements, was waving his arms, praying to it as if it was the sun.

The landlady noticed, while showing them the room, that they seemed rather distant. The man — and she had never seen a fatter one for his age — stood hat in hand, leaning on the post of the huge brass bed and looking out of the window across the marsh, the light catching on his gold-rimmed glasses and picking up his pearl-grey suit. The woman, dressed a little primly in a dark walking suit with a pull-on hat, paid no attention to the bed or the view, but was chiefly concerned to know whether the tasselled bell-pull beside the bed worked or not. On being told that it did, she turned her attention to the wallpaper, and did not speak again during the time it took for the landlady to conclude her agreement with the man and leave them.

Whether the old pig was hungry or not, or belching or yawning, or not, it still laughed at him. Whether or not it laughed at him he was still humiliated in the last degree.

The dank valley below was stopped with mist: the clearer slopes, if there were any, could not be seen. In all directions the ground was deep in pulpy vegetation, probably many feet deep and rooted only in a morass. There were no tracks, no rocks; the terrain seemed to shift in waves as it was looked at. The only colours in it all were the darker greens and the darker shades of rot; everything fresh was bedded on the wet decay of what preceded it. All of the growths seemed monstrous, and some were gigantic. From the middle of the valley floor a single huge tree, many-headed, thrust itself up through the mist, through tangles of vegetation that fouled its trunk, to divide into three or four great giraffe-necks, that stretched out many yards, quite bare except for the clotted growths, big as houses, they each carried at their ends. These were horned, bearded masses like pineapple-tops, bison heads, with sickly white trailers hanging down for yards under them.

Treadgold's electioneering speeches were wretched affairs. He spoke from the horse-drawn landau that belonged to his brother's brewery, and could hardly be heard. He was very tall and thin, with a drooping moustache and downcast eyes. He wore a long black coat, folded his hands in front of him where his belly would have been, and hung his head down sanctimoniously. It was tempting to imagine the horses spurred on, the carriage driving away and leaving him dangling, in just the same posture, from a noose.

At closing time, Lily Maskell and Mr Morris sang a duet, almost nose to nose, with their arms round each other's necks, Lily's coat open to show her high white satin blouse, little pearl studs at her ears and her black hair drawn in tight; Mr Morris not quite her size, crop-head held back to look through his little glasses and down his nose at Lily, who was almost bearing him back.

INTERIORS WITH VARIOUS FIGURES

1 *Experimenting*

Experimenting, experimenting,
             with long damp fingers twisting
             all the time and in the dusk
White like unlit electric bulbs she said
'This green goes with this purple,' the hands going,
The question pleased: 'Agree?'

Squatting beside a dark brown armchair just round from the
             fireplace, one hand on a coalscuttle the other
             prickling across the butchered remains of my hair,
I listen to the nylon snuffle in her poking hands,
Experimenting, experimenting.
'Old sexy-eyes,' is all I say.

So I have to put my face into her voice, a shiny baize-lined
             canister that says all round me, staring in:
'I've tried tonight. This place!' Experimenting. And I:
'The wind off the wallpaper blows your hair bigger.'

Growing annoyed, I think, she clouds over, reminds me she's a
             guest, first time here, a comparative stranger,
             however close; 'Doesn't *welcome* me.' She's not
             young, of course;
Trying it on, though, going on about the milk bottle, tableleg,
The little things. Oh, a laugh somewhere. More words.
She knows I don't *live* here.

Only a little twilight is left washing around outside, her unease
             interfering with it as I watch.
Silence. Maybe some conversation. I begin:
'Perhaps you've had a child secretly sometime?'

'Hm?' she says, closed up. The fingers start again, exploring up
             and down and prodding, smoothing. Carefully
She asks 'At least — why can't you have more walls?'
Really scared. I see she means it.

To comfort her I say how there's one wall each, they can't
                    outnumber us, walls, lucky to have the one with
                    the lightswitch, our situation's better than beyond
                    the backyard, where indeed the earth seems to stop
                    pretty abruptly and not restart;
Then she says, very finely:
'I can't look,' and 'Don't remind me,' and 'That blue gulf'.

So I ask her to let her fingers do the white things again and let
                    her eyes look and her hair blow bigger, all in the
                    dusk deeper and the coloured stuffs audible and
                    odorous;
But she shuts her eyes big and mutters:
                    'And when the moon with horror —
                    And when the moon with horror —
                    And when the moon with horror —'
So I say 'Comes blundering blind up the side tonight'.
She: 'We hear it bump and scrape'.
I: 'We hear it giggle'. Looks at me,
'And when the moon with horror,' she says.

Squatting beside a dark brown armchair just round from the
                    fireplace, one hand on a coalscuttle the other
                    prickling across the butchered remains of my hair,
'What have you been reading, then?' I ask her,
Experimenting, experimenting.

2   *The Small Room*

Why should I let him shave the hairs from me? I hardly know him.

Of all the rooms, this is a very small room.

I cannot tell if it was he who painted the doors this colour; himself
                    who lit the fire just before I arrived.

That bulb again. It has travelled even here.

In the corner, a cupboard where evidently a dog sleeps. The
                    preparations are slow.

He is allowed to buy the same sort of electricity as everybody else,
        but his shirt, his milk bottle, his electricity resemble
        one another more than they resemble others of
        their kind. A transformation at his door, at his
        voice, under his eye.

This will include me too; yet I hardly know him. Not well enough
        to be sure which excuses would make him let me
        go, now, at once.

Shave the hairs from my body. Which of us thought of this thing?

3   The Lampshade

It is globed
        and like white wax.
Someone left it
        on the table corner
under the lamp-holder with the stiff ring.

Across its curve
        a few red strands stick.
Just now she wrapped her hair around
to stage an interview with it.
Inside the hair.

Now, beside the cupboard,
        skirt pulled down,
she sits on the floor
watching me
        through brass eyes.

Thinking what she told the lampshade,
        what it volunteered,
the moist globe in her hair.

Soon she'll stand up.
She helps me make the bed and gets us
brownish wet food about this time.

The white globe stays with us
at mealtimes.

4 *The Steam Crane*

Before breakfast you drew down the blind.

Soon it will be afternoon outside. Hear the steam crane start up
          again

Deep in the world.

You sprawl with no shoes, wet with something from the floor
          you didn't see in the dark.

Black skirt. Black hair. Nothing troubles you, you big
          shadow. Much time has fallen away.

Wearing a blanket I sit in a hard armchair, a jug at my feet.

There's nothing I can give you as beautiful as the flowers on the
          wallpaper.

Under the wallpaper, plaster, bonded with black hairs.

5 *The Wrestler*

Stripped more or less, they wrestle among the furniture on his
          harsh green carpet;

This is their habit, the three of them, these winter mornings.

And this is my time for being with him afterwards. When I hear
          the others go downstairs I come to him,

Finding him spread squarely across a sofa, shirt and tie and brown
          suit pulled straight on again over his sweat.

He needs me there. Alone, he might drop the bottle and be upset;
          he might go down to thresh on the carpet again. I
          think this could happen.

But now he's still, only his fingers working through his stubble
          hair, suddenly across the face, down the bent nose:

The colours in his eyes have run together, and he stares up at the
unlit bulb that keeps constant distance from him as
he floats backwards to the ice.

He says nothing, though his mouth is open. Whisky is a fluid
squeezed out of damp ropes, wrung out of short
sweaty hair.

He's glad to have me just sitting.

She has left an empty glass, a cigarette butt in the fireplace and a
tissue here in the wastebasket;

The man an empty glass and the present of a cigar for him to
smoke after lunch, when the television sports shows
start.

Those two always choose the morning: a time when he's barely
civil.

The carpet straightens its pressed patches. Drifting back where he
sits he travels it like a cloud shadow, breathing
more gently.

The bull's eye in its jar of formalin, usually on the mantelpiece,
still sits out of harm's way on the cupboard-top.

6 *The Foyer*

The foyer's revolving doors are fixed open to let the dull heat
come and go;
This afternoon, old woman, the hotel extends a long way through
the streets outside: further than you've just been,
further than you can get.

Collapsed long-legged on a public armchair beside the doorway
under the lamp whose straight petals of orange
glass hide its bulbs,
You can't see the indoor buildings along the street;

You can see only me, roofed in with lassitude in the armchair
                    opposite,
Against the brown panelling, under the criss-crossed baize letter
                    board.

And not even that do you see, one hand spread like a handkerchief
                    over the middle of your face —
My hand feels cautiously across my summer haircut; my suit's too
                    big.

Your dress and cardigan, flowery and crisp, stand away from your
                    brown collapse and resignation like a borrowed
                    hospital bathrobe.

The heat flushes you in patches, the confinement takes your breath;

So many things are ochre and mahogany: the days, the flowers,
                    an attempt to look a dog in the eye;

This seems to be the place where they wrap us in paper and tie
                    us with string.

Though the windows are square and dingy here they're too big
                    for you, the ceilings too high to think about,

The doorway too lofty.

To cut any sort of figure going out, you'd have to let me carry you
                    through on my shoulders.

7 The Wrong Time

It's the wrong time; that makes it the wrong room:
I'm here, he's not; he was here, he will be.

Meanwhile, please use the place. It can use you,
Your scent, silk, clean lines, mouthwash conversation.
With him away it's sour and frowsty. You have
To swell your light to absorb the faint bulb, scuffed greenish
                    walls, breakfast wreckage,
Till the silk stitches hurt. You win. But the place contains me.

I'm not what you want. You're not what I want. What do you
            do with me?
Do you take me in, with the milk in the bottom of the bottle;
            dazzle me, with the grease spots, out of
            reckoning?

Or do you see round me, a man-shaped hole in the world?

Looking at you, I can't tell. You don't seem to find it hard,
            either way.

8   *Truants*

That huge stale smile you give —
Was it ever fresh?

Ancient sunlit afternoon
On a ground floor below street level.
The back window,
Curtained with dark chenille,
            spinach-coloured,
Gives on to a brick wall.

Poor quarters for you, old carcass,
But the cushions are fat over the springs;

You had plenty in that bottle, too.

For me, this is a truancy,
Five minutes' tram ride out of town
In the wrong direction:
I could feel trapped.

            You're different,
Truant entirely, inside your smile.

Two grown-up people.

9 The Arrival

You have entered, you have turned and closed the door, you have
          laid down a package wrapped in cloth as dull as
          your clothes and skin.

You have not looked at me, you have not looked for me, you have
          not expected me.

You busy yourself with the package, bending over it, your scuffed
          backside towards me.

You remain in the dirty shadow that edges the room, filtered
          through the fringed green lampshade; only here,
          just in front of me, does the light fall on the
          carpet.

You would think the light had eaten it away down to the threads.

You, being you, would expect the light to do a thing like that.

You wouldn't notice.

You might expect the faint smell of gas in here to have
          materialized into something like me. I want to go
          out.

You might notice my leaving. I shouldn't like that.

10 The Billiard Table

Morning. Eleven. The billiard table has been slept on.
A mess of sheets on the green baize.
Suggests a surgery without blood.

Starting the day shakily, you keep glancing at it
Till the tangle looks like abandoned grave-clothes.

And watching it from where I sit
I see it's the actual corpse, the patient dead under the anaesthetic,
A third party playing gooseberry, a pure stooge, the ghost of a
          paper bag;

Something that stopped in the night.

Have you ever felt
We've just been issued with each other
Like regulation lockers
And left to get on with it?

Nobody would expect
We'd fetch up in a place like this,
Making unscheduled things like what's on the table.

No longer part of us, it's still ours.

Bring the milk jug, and let's christen it.

As He Came Near Death

As he came near death things grew shallower for us:
We'd lost sleep and now sat muffled in the scent of tulips, the
            medical odours, and the street sounds going past,
            going away;
And he, too, slept little, the morphine and the pink light the
            curtains let through floating him with us,
So that he lay and was worked out on to the skin of his life and
            left there,
And we had to reach only a little way into the warm bed to scoop
            him up.

A few days, slow tumbling escalators of visitors and cheques, and
            something like popularity;
During this time somebody washed him in a soap called Narcissus
            and mounted him, frilled with satin, in a polished
            case.

Then the hole: this was a slot punched in a square of plastic grass
            rug, a slot lined with white polythene, floored with
            dyed green gravel.
The box lay in it; we rode in the black cars round a corner, got out
            into our coloured cars and dispersed in easy stages.

After a time the grave got up and went away.

Colour Supplement Pages

Two dimensions. Diptych divided by a house wall seen end-on.

To the right, the street, travelling rapidly towards you and away,
                        folding itself into and out of itself and crouching
                        as low as it can get,
A street with concrete slabs, stuck leaves from lime trees beside
                        the river,
Small billboards screening a lorry park, puddles in asphalt,
                        punch-marks of heel and ferrule, people with pink
                        wrinkles, plimsolls and moist jaws,
A barrow with cream wheels, a crushproofed old woman, a
                        showerproofed detective, a sunblown blonde;
Pale strides, advancing smiles, men crumpled together, blue
                        shadows on working shirts, wet baby mouths and
                        scaffolding.

Through the wall, raised a little and remaining still, the room,
Where a woman, tall and rich-bodied, with thick dusky skin and
                        matt black hair falling to the pants that are all she
                        wears,
Stands glaring vaguely at a television screen:
A brown longhaired dachshund
Raises its head from dribbling on a cushion, and watches her for
                        a while.

— Frame it in glitter, or dirty reseda.

After Working

I like being tired,
to go downhill from waking
late in the day
when the clay hours
have mostly crossed the town
and sails smack on the reservoir
bright and cold;

I squat there by the reeds
in dusty grass near earth
stamped to a zoo patch
fed with dog dung,
and where swifts
flick sooty feathers along the water
agape for flies.

The thoughts I'm used to meeting
at head-height when I walk or drive
get lost here in the petrol haze
that calms the elm-tops
over the sunset shadows I sit among;

and I watch the sails,
the brick dam,
the far buildings brighten,
pulled into light,
sharp edges and transient,
painful to see:

signal to leave looking and
shaded, to fall away
lower than dulled water reaches,
still breathing the dog odour
of water, new flats, suburb trees,
into the half light of a night garage
without a floor,

then down its concrete stems,
shaded as I go down
past slack and soundless
shores of what might be other
scummed waters,
to oil-marked asphalt
and, in the darkness, to a sort of grass.

Seven Attempted Moves

If the night were not so dark
        this would be seen
Deep red,
        the last red before black.
Beside the soft earth steps
        a wall of heaped stones
Breathes and
        flowers
  and breathes.

      *

A cast concrete basin
        with a hole in the bottom
Empty but for
        a drift of black grit
Some feathers some hair
        some grey paper.
Nothing else for the puzzled face to see.

      *

Crisis —
        a man should be able
To hope for a well made crisis,
Something to brace against.

But see it come in rapidly and mean
        along some corridor
In a pauperous civic Office.

      *

Under the portico
Huge-winged shadows
        hang
Brown, with a scent
        of powdered leather.
Up the steps
        into this
Depth. Recession.
Promise of star-scratched dark.

Then put your ear to the door;
          listen
As in a shell
          to the traffic
Slithering along behind it.

                    *

Here are the schoolroom chairs on which
          the ministers, in the playground,
Sat to be shot.
          Four chairs; the property
Of the Department of Education;
Stolen
          the same night
By this souvenir hunter
          with his respect for neither side —
Just for things happening;
          then sought in vain
And after a long while
          written off the books
Of the Department of Education.

                    *

Bright birchleaves, luminous and orange,
Stick after six months to the street,
          trodden down;
Now, as at every minute, perfect.

                    *

It is a shame. There is
          nowhere to go.
Doors into further in
          lead out already
To new gardens
Small enough for pets' droppings
          quickly to cover:
Ceilings
          too soon, steps curtailed;
The minibed; minibath;
          and jammed close
   the minican.

Confinement,
            shortness of breath.
Only a state of mind.
            And
Statues of it built everywhere.

The Thing About Joe Sullivan

The pianist Joe Sullivan,
jamming sound against idea

hard as it can go
florid and dangerous

slams at the beat, or hovers,
drumming, along its spikes;

in his time almost the only
one of them to ignore

the chance of easing down,
walking it leisurely,

he'll strut, with gambling shapes,
underpinning by James P.,

amble, and stride over
gulfs of his own leaving, perilously

toppling octaves down to where
the chords grow fat again

and ride hard-edged, most lucidly
voiced, and in good inversions even when

the piano seems at risk of being
hammered the next second into scrap.

For all that, he won't swing
like all the others;

disregards mere continuity,
the snakecharming business,

the 'masturbator's rhythm'
under the long variations:

Sullivan can gut a sequence
in one chorus —

— approach, development, climax, discard —
and sound magnanimous.

The mannerism of intensity
often with him seems true,

too much to be said, the mood
pressing in right at the start, then

running among stock forms
that could play themselves

and moving there with such
quickness of intellect

that shapes flaw and fuse,
altering without much sign,

concentration
so wrapped up in thoroughness

it can sound bluff, bustling,
just big-handed stuff —

belied by what drives him in
to make rigid, display,

shout and abscond, rather
than just let it come, let it go —

And that thing is his mood:
a feeling violent and ordinary

that runs in among standard forms so
wrapped up in clarity

that fingers following his
through figures that sound obvious

find corners everywhere,
marks of invention, wakefulness;

the rapid and perverse
tracks that ordinary feelings

make when they get driven
hard enough against time.

For Realism

For 'realism':
the sight of Lucas's
lamp factory on a summer night;
a shift coming off about nine,
pale light, dispersing,
runnels of people chased,
by pavements drying off
quickly after them,
away among the wrinkled brown houses
where there are cracks for them to go;

sometimes, at the corner of Farm and Wheeler Streets,
standing in that stained, half-deserted place

— pale light for staring up
four floors high
through the blind window walls
of a hall of engines,
shady humps left alone,
no lights on in there
except the sky —

there presses in
— and not as conscience —
what concentrates down in the warm hollow:

plenty of life there still,
the foodshops open late, and people
going about constantly, but not far;

there's a man in a blue suit
facing into a corner,
straddling to keep his shoes dry;
women step, talking, over the stream,
and when the men going by call out, he answers.

Above, dignity. A new precinct
comes over the scraped hill,
flats on the ridge get the last light.

Down Wheeler Street, the lamps
already gone, the windows have
lake stretches of silver
gashed out of tea green shadows,
the after-images of brickwork.

A conscience
builds, late, on the ridge. A realism
tries to record, before they're gone,
what silver filth these drains have run.

At No Distance

Antiphon: Two Parts for the Same Voice

It has disintegrated,

> the world across the bay
> silk dress with crayons
> cistern paint;
> > desired
> and remembered things

And everybody's like me

> Over the bay
> at evening raised steam
> silently in a far pool
> sun on the orange
> clouded water

Central to ourselves

        the dress wrinkling,
        somebody wearing it
        lying prone across
        stretched to collect
        from the floor the scattered

There,
Under that labyrinth
Of roofs they're the same —

They see
Lead paint fresh-run
On rivets, blobbed
Paint-skin to be broken
By anybody,
Smelled on the fingernail

        Enamel, coating
        the brown pencil crayon,
        nicked through
        to the wood

        fingers smell of it
        a woman's smile like

Lake, rocking high-bodied
Under wood chips,
Cold boats bucking

        and streaks of wax pink
        across the penknife blade
        brown and viridian

        sliced through then
        to the lead colours.

Some things always close:
Giant iron
Casts, no dream, and
Gantries I have made into
Arches that bear sense
Are no distance from me

                    the black leather
                    male shoe

Or from other men

                    worker-prison

Old Krupp
In the fireproof Villa Hügel
Moving from room to room
Ceaselessly as he felt
His presence foul the air

                    the colossal woods
                    banked up behind the lake
                    sunlit;

                    there are things men own
                    ample enough
                    to weigh heavy
                    in longing's belly

The dress, cold
Wrinkles under the breasts,
No distance
From freckled skin beneath

                    though smelling of himself
                    must have thought

                    freedom from his own foulness
                    a metaphor: mountain air,
                    lake water, gunsmoke

Old Krupp, fearing
His own fires, his own lead

                    in bad weather
                    behind the windows
                    a woman gathers herself in
                    close to herself,
                    whoever longs for her

Whoever owns.

Wedgwood, too,
Built his dwelling
A slate blue barrack
From where he could look down
And watch Etruria
Making it

> there's
> our bright cloud again
> swelling far off

Glancing from the works

Up across the meadow and the water
Showed him, distanced, his
Rectangular domestic peace

> what the imagination calls
> power: to own what's longed for
> and play both ways —

Grotesque hermaphrodite,
Groin without orifice,
That can own what, and whom,
It doesn't even desire —

> crayons
> sharpening to brown
> pink and viridian,
> smelling of coloured
> leads and cut wood

Landing stages across the bay,

Nondescript spot —

> unfastening down the back
> and crumpled more
> fallen away round
> freckled skin —

Yes, conscious of itself
Centred like all of us;
Even like Krupp.

                    From far in memory
                    the red cistern
                    freshly painted

Someone else had done it
One sweaty day

                    that part
                    beyond all distance

It was already done
When I found it

                    one various world
                    beckoning infinitely
                    to make me dream

to make me do;

Or many worlds

                    someone bought
                    the forest behind the lake

Collide, precipitate

                    cold afternoon
                    paper and crayons

Making one
Various world

                    the view across the bay

Looked at
From anywhere on it.

The Memorial Fountain

The fountain plays
        through summer dusk in gaunt shadows,
black constructions
        against a late clear sky,
water in the basin
        where the column falls
    shaking,
rapid and wild,
        in cross-waves, in back-waves,
    the light glinting and blue,
as in a wind
        though there is none,
                Harsh

skyline!
        Far-off scaffolding
bitten against the air.

           Sombre mood
in the presence of things,
        no matter what things;
respectful sepia.

        This scene:
        people on the public seats
        embedded in it, darkening
        intelligences of what's visible;
        private, given over, all of them —

Many scenes.

Still sombre.

As for the fountain:
        nothing in the describing
beyond what shows
        for anyone;
    above all
no 'atmosphere'.
        It's like this often —
I don't exaggerate.

And the scene?
a thirty-five-year-old man,
poet,
                by temper, realist,
watching a fountain
and the figures round it
in garish twilight,
                working
to distinguish an event
from an opinion;
                this man,
intent and comfortable —

Romantic notion.

Report on August

How do I sleep? Well, but
the dreams are bad:

filled with accusations
small but just.

These slack summer dawns
that fail of sunrise

there's a relief at falling
awake and into comfort,

becoming once again
four people, watching

from pillow level
my boys' khaki heads bustle about.

Over breakfast I see,
staring at the garden,

how the times have fed:
under heavy leaves and low sky

in profile the bold woodpigeon
walks the lawn.

Beats of a shadowy fanblade
tick through from behind,

time going; ignored,
nobody measuring time, so much

constant, the weather unchanging,
the work I do filling days

so that they seem one day,
a firm framework, made

of the window where I sit
(or lie, slumped, feet on the desk,

waved to by passers-by
like a paraplegic)

a window-shaped guise of myself
that holds what few events come round

like slides, and in what seems
capricious sequence.

From an English Sensibility

There's enough wind
to rock the flower-heads
enough sun
to print their shadows
on the creosoted rail.

Already
this light shaking-up
rouses the traffic noise
out of a slurred riverbed
and lifts voices
as of battered aluminium cowls
toppling up;
black
drive chains racking the hot tiles.

Out in the cokehouse
cobweb
a dark mat
draped on the rubble in a corner
muffled
with a fog of glittering dust
that shakes
captive
in the sunlight
over pitted silver-grey
ghost shapes that shine through.

In Touch

I took down *Pictures from Brueghel*
                              to see what ways Doc
Williams had of taking off
                    into a poem

                              a strong
                    odour of currants
rose from the pages —

well that was one way.

Studies

1
Convalescent
to get along the shop fronts
far as a butcher's awning.
In the wind and sun, pause.

Quiet afternoon
to be powerless
held up by clothes
muffled
from the empty bright world
unrecognized
near home.

Library books
mostly about mountains
Kangchenjunga
in the silver crust
hanging on the gulf beyond Darjeeling;
Godwin Austen in the silver crust.

Deep sky between the fragile
clouds
backing the library weathervane
set on a cupola of green.

2
Civic water
with lilies and flies.

Men extend a trellis. It
opens across a flowerbed
to enclose —

Boats recede
shouts
fade among the olive
green expanses.
The sense of it all slides
faint-tasting
outwards from pain
low in the body
in the registers of the body,
among the all-but-kicked-against
fencing of split birch-logs,
the pergolas —

Even these sometimes leave it,
the discomfort
to be without present image:
body
trying to walk itself off
to a memory of health;

or at times
hope
when the water's seen to move;
a fleck of light on the ripple.

Quarry Hills

Tail lights turn off the quarry road
into comfort
from desolation there
to new houses of cinder brick
with cemetery gardens.

Nothing much moves.

Only a feeling that strokes of shadow
flicker down the hill
walking,
as strokes of rain fall on a field,
down across the gutters
constantly through the shallow dark.

Hawthorns thick with lorry dust,
trackways slimed with it;
in forecourts
scents of hot engines die away.

Strawberry lampshades.

All the machinery
rickety and ageing;
the hills dwarfish,
almost eaten through with roads.

White Cloud, White Blossom

Sandy wall and trees
locked to the sky

small country
buckled iron fences
ducking in the grass

leaf print shadows,
fingers on a soft big belly
out in the sun

old bricks patch the mudhole
where cows come through the hedge,
dead swags of bramble shake on the wire,
under the elders
dry twig beds lie

steam from a mash boiler
blowing down through the wood

white cloud, white blossom

Three Ceremonial Poems

1
Metallic sheath
derelict, that resounds

Laurel bars, enamelled
with laurels, the bronze
on matted hair, blades
designed on guns, sharp leaf

Blue brick framing a door,
further, medallion,
fallen;
gallon can,
shield's mark on water,
the rusty flow

Fold upon fold, so stiff
and out of the clay-track
out of it
to walk and bed on flags,
damp bundles,
padding the stone

Salt crust,
crystals break out and part,
open on flesh; in frost
leaf, sharp curled,
like leather, sticking fast
where wheels were down

Sunset and dusk mark it,
and the pulling away,
fold on fold,
from matted places

Covering of the stacks; by twilight
live mask plated over

Warrior, the stopped man.

2
Absolute
Pity
Advancing

Out from the grove
Of parsley elms
On those that wait
Staining
Suit-knees
With grass juice.

In the concrete rank, a panther cage.
The panther hates, at morning and at midday;
Lies in the dust and stifles the sunbeams.

— But turned inward,
Studying one's very own:
This curious corner—

A urine-softened wall
Meets an impervious hard one;

Clay, cut like butter,
Drags at the trowel,
In cold sweat subsides;

And golden drops
Shake, and fling
From the body, brightness

Trembles the window,
Dark strands
Lick outward down the arms.

Fading
Short
And sudden when it comes,
White flash
lost under the anvil cloud;
Dark dust of shame

Raining down
Deep in the brickwork,
Changes its face;

And falling
In the open,
Left out among the trees.

3
Oh yellow head,
Crust of deception;
Pale over wheatfields, the sun.

Suppose —

Suppose that once in a while
It still works, just as it used to.

Somebody unwraps it among the teacups,
Curtained from street flashes
By afternoon clatter,
A crowd of faces and feet,
That sort of thing;

Opens it, finds a poem —
The old flat arrangement,
Dry track of half a voice —
And lets it drift on his own thoughts,
Like a simile.

As a mirror, held to face another,
Deepens it with recessions
This used idea, abandoned
And pinched up into caricature,
Monitors and shakes the new.

Between them, a guttering freedom,
Just enough light to ask questions by:

Why Aleksandr Blok, the beautiful,
Dealt out humbug,
Still made sense —

The Making of the Book

'Let the Blurb be strong,
modest, and true.
Build it to take a belting;
they'll pick on that.

Then choose your second gang —
the first, led by your publisher,
you already belong to,
its membership involuntary, if free —

for the other, set up an interesting
tension between the acknowledgements
and the resemblances; but in the photograph keep
the cut of your moustache equivocal.

Write your own warrant. Make plain
in idiot-sized letters
for which of the others you'll take the blame — Yes:
it's *necessary* to belong;

several allegiances
are laid out for you ready.
And remember, though you're only a poet,
there's somebody, somewhere, whose patience

it falls to *you* finally to exhaust.
For poetry, we have to take it, is essential,
though menial; its purpose
constantly to set up little enmities.

Faction makes a reciprocal
to-and-fro of the simplest sort — and characterless
but for an "aesthetic" variable,
inaudible to all but the players.

And this little mindless motion,
that nobody but the selfless and Schooled-
for-Service would ever stoop to,
drives the Society.

It's a long story:
but the minuscule dialectic,
tick-tacking away, no more than notional,
in obscure columns,

at length transmits itself
mysteriously through Education —
which pays off the poets too,
one way and another —

out beyond Government,
past Control and Commodity
even to the hollowness
of the seventeenth percentile, the outermost
reaches of the responsible.

If the reviewers fall idle, everybody drops dead:
it's as simple as that.

— *Go, little book.'*

Continuity

In a covered way along the outside of the building,
A glazed lean-to with the panes painted over,
There's nothing but the light, falling across the quarries
From the lamp in the wall above.

Purpose? No purpose. Apparitions? None.
Where the lamplight peters out across the yard
Two shallow plant pots swim the night.

There is smoke folded into the waters,
The fish-trap gives the waters form,
Minimal form, drawn on the current unattended,
The lure and the check. So much free water.

Long clouds are lined.
Over the level-crossing ramp.
Faces behind car-glass painted with reflection,
Pressed in the seats, warm bodies;
And the exhausts patter on the dirt
Stained through with oils, sterile with gases.

When they pull away across the ballast
Dirty Nature claims her own.

Tongues of grease in rings of light.

The towns are endless as the waters are.

Glenthorne Poems

for Ursula and Ben Halliday

1
Straight into the sea fog
the descent
drops
red track
with sudden angles
turn below turn
into the white

Face
over the drop
the car blank
face of the descent

From the moor road
red track seaward
skirting a crater rim
charred bracken black
brimming the fog

Falls from the world
doubles down
into white

At sudden turns to meet
treetops below
hillsides of rhododendron
drifting
higher into the white

A different depth
opening
past savage gateposts

the hills' cleft
narrowing to a floor

Towers in the woods
and over the last pines
the sky wrinkled
call it sea

A different depth
with light soft
on roof leads

No sound but breakers
under the cliff
the hills deep
the sea standing
into the cloud

2
I have slept shallow a long while
living where car park lamps
neutralize the night
to a damping of action circuits

rooms to street annexes

and sleep to readiness

3
Gone down
from the upper air

Sunk under the hill
as in my own
sleeping body

Possessing nothing
warming the spring
with paraffin and pear wood

After midnight
when the generator cuts out
listening to the trees

4
Real things move
as if they were free

Pillars of smoke
rose coloured
white
smoke fans
flat like burner flames
suddenly displayed
on silver squares

All that
is Glamorgan

Celestial Aberthaw
breaking above the haze
a dozen miles across channel

Clear day
lighting the woods
here on the headlands

And as if they were free
patterns the breeze inshore
makes on the blue

far below
move under the trunk of a pine
down on the cliff edge

Eastward in the sea
Steep Holm
Flat Holm
islands towards Severn
keep their distance

Then dulling of the sky
darkens the Welsh coast
to what it is

As if free
it pushes closer

5
At sunset over the water
the nondescript cloud
builds up and breaks
in dirty dramas across the sky

With colours from clinker beds
brilliant in paradise rims
or washed wide

Sun dazzles along the waves
and slaty shoals of low water

Strikes up the cliff
in under the dark of the bushes
with tangles of burning wire

To chance on a gold thrush

6
The hills lie thick
under a three-quarter moon

The bank falls steeply
from the big oak

The oak's bare branches
fill the sky

There are stars in it
and other lights

In red stacks and white lines
mark out the sea horizon

7
Freedoms
out of contrivance

A stillness
breaking at the edges
to paradigms of freedom

This stillness I contrive
branches low under the mass
and runs for the dark

Weight to be shifted

I have
a worn black pebble

hangar-shaped
and traced over with veins
that stand to the touch

Once
I watched it ingest
a violin concerto
of Bartók entire

8
A steep turn down
to a water butt

A wall
of Mexican Orange Blossom
its shiny three-lobed leaves
scenting the hollow

An iron roof
cinnamon with rust

Beyond it and the long
wall of the garden
the cold sea runs

9
Walking at dusk often
twelve years back

On pavements of the hill
where we were neighbours

worn volcanic knob
over the Teign valley
planted with roofs
and a few pines

I'd hear the pumps
chuck and clink
out in the clay workings
weed islands
on white
lakes of slurry
over to marsh meadows
where the single railway track
ran past plantation belts
up into the foothills of the moor
where there were shadows
darkening red
mile after mile in the folds
under Haytor

And finally I'd be left
With hardly a street lamp under me

Walking on the dark

10
From the wrecked byre
with dungheap
and periwinkle bed
two narrow fields
fall between woods to the cliff

To one side the stream
cuts under them faster
twisting through water gardens
to a rough cleft where sea shows
then a fall

On the other
the old beach path
mossy and walled in
turns off under the wood

I see the trees hung over it
coming into leaf
a line of them

This is how it was
when first I started with poetry

They are already
three-parts idea

The Six Deliberate Acts

1
Today
let us think
and the smoke rise from the pan

Come in together
with what's behind us
leaning at our backs

Gentle eye can see
débris stuck to the burner
shoot comet tails in gas flame
and crumble away red

Paper resembles skin

On skin comes linen

Pale walls without much echo
out along the passage

2
Emptying his bowel
he sat in a small outside toilet
among dustbins and fire escapes
at the back of the new shops

His turds were dry enough to rattle
but no kind of trouble

The shops were high on a shoulder

built up from dumped ash

and the view from the jakes door
he held open and peered round
dropped over the concrete walkway
straight into the dip

where there was a field of kale
stinking and muddy

Past that the double road
went curving off below

There was a row of leafless poplars
and beyond them hoardings

Across the scene
the railway
went out along a cutting
bridge after bridge after bridge

And clanking into space
hoppers from the brickworks
swung out over the road

on a cable track
down into the sidings

Some way off
on a hillside
the Rehabilitation Centre
had its lights on in the afternoon
and over against the skyline
under blue drops of early streetlamps
a housing estate
filled in the outlines of a farm

Squatting up over the land
backed by the betting and pie shops
hearing the launderette
it was cold with the door open
but high

3
To see the film you buy a ticket

Some say it's just a room you pass into
with air outside

Some say the ticket makes it dark

Got from a woman behind plate glass
with a stokehole at the bottom for the money

The coins come into the light
and through the glass

They pass to the woman on the stool

Shoe grease and touch of nickel
smoky mirrors in anterooms

The edges of the dark are scuffed

But clean as a whistle
the black middle of the pit

4
The only target
is the eyeball

Isolated in its fringe
staring concentric at itself
in a round glass
on a peppermint wall

Copper hair
quivering in the light

she pops the first lens in

Things will move on

The glass when it goes
leaving a dirt-rimmed paler disc

a flat eye

5
Emptying his bowel
in the toilet behind the shops
he looked out across the landscape

A plain ground as if white
stretching away nondescript
set out with tracks
and heaps of things
all made of marking

The more he looked
the more he saw

Hedge-marks
smoke-marks
roof-marks

Until it balanced
and tipped the flush
just as he reached for it

6
He is going to the woman
and he is right

Well past the middle of his age
the sun paler than lemon

Dampness in the flat gardens

He starts to go to the woman
when he starts his journey

There are buses
a bus ticket

There is trajectory of the sun

He will arrive with himself

METAMORPHOSES

1

She sleeps, in the day, in the silence. Where there is light, but little else: the white covers, the pillow, her head with its ordinary hair, her forearm dark over the sheet.

She sleeps and it is hardly a mark on the stillness; that she should have moved to be there, that she should be moving now across her sleep as the window where the light comes in passes across the day.

Her warmth is in the shadows of the bed, and the bed has few shadows, the sky is smoked with a little cloud, there are fish-trails high in the air. Her sleep rides on the silence, it is an open mouth travelling backward on moving waves.

Mouth open across the water, the knees loosened in sleep; dusks of the body shadowed around the room. In the light from the window there is the thought of a beat, a flicker, an alternation of aspect from the outside to the inside of the glass. The light is going deep under her.

Enough depth. To clear and come free. There is no taste in the water, there are no edges under it: falling away, the soft mumbled hollows and mounds of marble, veined with brown, a lobby floor gone down into the descending levels of a sea-basin. The sleep comes naked.

Rising through the clear fluid, making their own way, the dragging wisps of brown that were secret hairs or the frame of a print on the wall. And light that cracks into the bubbles near the surface, lighting them like varnish bubbles, breaking them into the silent space between the surface and the curved roof, threaded with moving reflections of water light.

Water lights crossing and combining endlessly over the inward membrane of the roof, rising in a curve, almost a cone, to the round lantern with its dirty panes. The water lights beating silently under the steep slates of the case, under the painted frames of the lantern's windows and the domed lead cap, holding into the sky a two-foot fluted spike.

2

The cat's glinting face as it stared up between its paws from the odd soft position, near-supine, into which the other cat had rolled it. The other cat was already indifferent, turned away to lick its upthrust leg, but this cat, for the moment, had no next thing.

The cat's glinting face, all mask, no signal, was an old face: on a man it would have been frozen, the defiance of something contemptible. It was framed in soft paws. It stared indiscriminately up at the lights. The cat's eyes, further away than eyes look, more distance in them, no cat. Running into the distance there's a dull aluminium strip of road, tall skies, flat horizons, with scattered elms and poplars picked out in colour by the sun.

All the green fields are cold, the bright afternoon deserted. Faces look out of the cars that go by; that is what they do, those faces. There is a tower among the trees, a white drum on legs, and a road turns off beside it, sweeping down to a cinder patch by the river where the field-tracks join and cars can park. A path, much mauled and trodden, leads through the elders, and at one place, where it crosses a marshy dip, a sheet of corrugated iron has been wedged, balanced on a springy root and half earthed over.

3

How does he come to be wearing that suit, clay-coloured, with a hang-off jacket and flapping trousers that make him seem to jerk? He's making for the ferry; no he's not. He stands a while and goes somewhere else.

A man among the puddles with his shoes on the pavement and his head in the clear air, his nervous system shrouded in loose, dried-looking clothes. His trajectory leads these arrangements he has to a pause, then takes them past it. While he is there and after he's gone, the shut car-park kiosk remains unwaveringly present.

For a few seconds in the centre of a rigidly composed scene, its elements stopped in the act of crossing from left to right or right to left, there is, maybe in the name of freedom, maybe in the name of compulsion, an unidentified capering, that leaves no trace after it has gone. The pavement gives place immediately to the air above it; there was the sign of a thing like a man in the air, an eddy across the scene.

No system describes the world. The figures moving in the background stop and wait in mid-step, the sound-track cuts out: the projector motor runs on, the beam doesn't waver. Among the whites and greys of the picture a golden shade is born, in the quiet, rippling slowly, knotting itself and suddenly swelling into a cauliflower head, amber and cream cumulus outlined in blistering magenta, erupting out of itself and filling the screen before shrivelling off upwards to leave a blank screen and a stink of fire.

4

Red beans in to soak. A thickness of them, almost brimming the glass basin, swelling and softening together, the colour of their husks draining out to a fog of blood in the water.

The mass of things, indistinguishable one from another, loosing their qualities into the common cloud, their depth squashed by the refraction and obscured in the stain, forms pushed out of line. Five beans down it may be different.

Down in the levels, it's possible to think outward to the edge; with a face to the light, there's no looking out, only hunching before the erosion. Back!

In the midst is neither upward nor downward, head nor foot has precedence or order. Curved belly rises above, warm and shining, its navel out on the surface with the vestige of a lip. One eye is enough, to distinguish shape from shadow, paired eyes would fix too much. To be fixed in the midst is suffocation.

So, in the thick of the world, watching the moon whiten the bedroom floor and drag the print of the window nets higher and higher across the wall; thinking how the world would have had a different history if there had always been not one moon in the sky, but a close-set pair. It is said there are two breasts; it is said there are two sexes. That's as may be.

Out in the moonlight is a short street with only one side; houses on it, and walled forecourts. Over the way is a white pavement, and a blackness where the hill falls away. The blackness goes grey with looking, and the valley is full of shapes.

5

Crumpled clothes come slowly off and fall to make a heap on the carpet. In this daylight, the nakedness is theirs: pulling his shirt off, letting his belt snake out of its loops and drop, he is closing inward, tired, to his own body's shadows, while the white underwear faces up to the light from collapse.

The clothes are falling to make a different naked body, loosing the bands that wrap him in; he draws himself into his shade, everything is outside him; the nakedness of his belly crumples out on his shirt.

Seams and bands that contained him are falling away, in shapes to suit themselves, different shadows, different surfaces. Not dispersed from one another, they move into a scheme that frees him and does without him.

It frees him so he can feel it go from him once the clothes have gone, as if a membrane webbed over with straplike lines was parting across his belly, his breast, the fronts of his thighs, winding off across the daylight, that comes to back it with chalky aquamarine, and slowly turning off-centre as the twisted straps float out, and show themselves in blotted-down tapestry blues and reds.

They trail in the greeny blue, where there are dark dried fragments of clover: a stone coping at the edge of the blue has the colour of mushrooms cooled and cut after scalding till they flush, a matt fair skin scattered with wide freckles, thickly over the sun-reddened base of the throat and the stretch beneath, but more and more rarely as it rises pale and smooth to the nipple.

Critics Can Bleed

His passion for the books
warms them

His warmth
softens them

Then his wits
work them.

Matrix

1
So, the water gate
starts out of the lake
or inlet

and petrol-blue
around its pier-timbers
the waters, after sundown,
draw their lymphatic currents by,

circling the clutch of islands,
or segmented Isle;
brimming on the shores.

Alone, it could be a house,
chin on the water,
hat-roofed,
but rising behind it
there is more.

So it gives footing
first on to rock

where channels cut shadow:
or was it the last thing built
of all the provisions
in the pattern —
the one lacking purchase
and pushed on to the waves?

Domestic-secretive:
secretive-institutional.

One of the smallest.

2
Eight or nine yards
of offered crossing:
outcrop, with pushed-up strata.

Levelled, the chasms filled,
pinnacles snagged off,
skimmed with a surfacing
that cakes into a path;

not altogether closing
over patches of rock-knuckle
and the backs of waterpipes.

3
There is a place of cypress fingers
thrusting from cypress mounds.

Low walls
built with old money.
Turf and aubrietia
painted around the stones.

This is where the dead
are still supposed to make
their disappearance:

but always the same dead
seem to be walking.

Spectres of respect.

The slopes that overlook the island,
or nest of islets,
light up with coloured squares
in among the elms:

separate plots of twilight
running the same destiny:
the boat, with muffled oars,
the hooded figure, B.D.,
its hand upraised,

a small classical temple
of the Lutheran cast
off to one side of the walk
on a knoll;

The mountains of the cypresses
are the real dark of the path,
humping their way higher;

always a dark like that,
set off by some artifice,

growing.

4
In long shapes over the channel's
fading blue,
footbridge slats,

with slatted sounds of crossing:
the shadows rattle on the blue
trough-water in the rock

where nobody lies,
nobody rests, at present.

5
Sugar flowers
on a windowsill
in the dark,
Saintpaulias in sugar,

pink, and flock violet:
everything eclectic,
building in deeper.

Spiralling from the shore
in blocked paths and on dropped
or jutting levels,
the stations of the thing
face over one another,

inseparable, interfolded,
wall-face into roof-angle,
catwalks over the garden-clefts
hung with valerian:

variants on cochlea,
invisibly thin
stone ear in the sky;

on how to get down
in through the horn
of the gold snail-shell
and not grow small.

One way and another
inward from the shore,
the narrow ground repeats itself:

trenches of silver water pierce
the island-cluster
or do no such thing:

it is all or nothing there.

6
Religious garden
dropping between two gables
ravine of black silk
twined with a cataract
and bearded flowers
their padded scents
coming as if through gauzes
printed with petals
in a draper's.

7
Ocean-lights go flashing
over the skeined roof of the sea:

miles of deep water,
northerly, heard in the dark
island-hole;

surface that races
under the wind;

that will not break;

whose lights sink,
distort, endlessly turn again
to the wave-glass,
glittering up out of the weight,

Fabergé medulla.

quartz-fires
in the under-vault,
scattered through softer globes,
their spiral trails
chasing and merging,

to form an uneasy field,
its charges drawing
the dwindling emerald waters
densely in,

as if, past all extension,
to the devouring drop.

8
Blood-red and blue glass
stain the air thinly
on a deep turn:

maybe it is approach —

maybe *kitsch*,
like so many orchids, or

the Night-Blowing Cereus,
by God, not Reinagle:

If it is approach,
it has come already;

if *kitsch* —
it turns to approach.

9
Doctor Meinière —
the cabin has to tilt
here most of all
planted on earth
with clematis at the window,

close under which the corner,
with moonlit carpet, drops
like drawers,
and goes on down;

the floorboards opposite
heave up, keep coming,
silent dog attack —
spreadeagle as the floor
becomes a roof
to slide off —
the whole room
falling betrayals
always the same way,
insatiable axis.

Afterwards
when the head turns
his liquid contents follow,
heavily drifting;
slow to a near-stop.

Somewhere in all the whirl,
exposure,

visions of the rock, Doctor,
bare and brown

thrusting a back up
with elephant-hide fissures;

dispensing
with sheets of grass and sky.

10
There's a time, finally,
when it doesn't matter
that the rings of the eye
slacken, and won't mark
so many differences:

whether the water-lilies
are blue, or the water,
or the sky in the water,

or the ocean-levels
through the confused floor
of the garden channel;

long white and green
ravels in the blue
tensioned over the shimmering
chalky surface;

net of fine hairs,
some grizzle, or black,
the weave on flank and belly,
delicate envelope,
whole surface on the move

filtering currents

tangled with trailers
of sky, and maybe lilies.

Correspondence

for Tom Phillips

I have a book, it has grown
from spots in the back cover
right on up
through the bank
to the surface of Page One

Words are lying there
arrived
Nowhere to go
but off the edges

Sooner or later
all things
have to be obeyed

— Caught sight of myself
in the monitor

The world looked like itself
I looked like it too
not like me
as if I was
solid or something

— Solid pushed through the curtains into the grove
for the thousandth time

His world

huddling him
front and back

It's not hard to look busy
from behind

— For *him* the cleared smoke
came back with its warrant
and back with its warrant

For him the sky inched open
the hewn beams rolled

— Far away now
from my home town
down in the monitor

Poem

The small
poem
the unit of feeling

Pretty red mouth
blotted and
asking why

Here is your photograph
It is a square
view of the air of things
one certain hour —

when just in the background
a green engine goes mad in a tree
to the end of time, pretty mouth.

The Sky, the Sea

The sky, the sea
and the beach:

at evening
there is a man left
on the glistening fade,

heading for the beach house
under a wind
that scrubs all sounds away.

There is a driftwood fire in the hearth,
the smoke pulls down on the roof;
there are two of them there,
it is simple;
and into the dunes
their window faces its white eye.

It is simple: the house
in the eye of the wind —
all else is ground, the long levels
sustain it, the small
knot in a grain washed clean
to silken grey —

All of it is plain:
the house also featureless, a Puritan
bungalow, bleached but strong;
and beyond what has to be done
there is nothing; the dusk
free to come down,
filled with cities of division.

From the 'Town Guide'

Out in the air, the statue
gets cold. It needs a coat.

The coat must have a face on top
to squint for dandruff on the shoulder.

It always did have trousers,
Remember? And a wife.

— She was a raver, great big
wardroby body. Insatiable. Still is.

She drives a car like that one
by the Conveniences. His epitaph

Stands all about. But on his plinth
read simply: 'The Unknown Alderman'

The Least

The least, the meanest,
goes down to less;
there's never an end.

And you can learn
looking for less
and again, less;
your eyes don't get sharper.

For there is less
eyesight;
and no end to that.

There is you,
there is less-you:
the merest trace —
less-eyes will find it.

Artists, Providers, Places to Go

The little figures in the architect's drawing
the sleep of reason begets
little figures.

Nose the car up through the ramps
into a bay, and leave it,
keys in the dash by regulation —
cost-effective:
come back and find it gone,
you got free parking.

The concrete multi-tiers
on the high-rise estate
hold everybody's wagon.
Only they don't. What's left there
their kids tear apart, Monkeyville —
anybody in their right mind would have known it.

Next, the Adventure Playground.

Next, celibate adult males
shipped in from the Homelands for work
sleep on long shelves of concrete,
Unity of Habitation. No damp.

— for that drawing, reduce
Sleepers in the Underground to cosiness,
consider the blanket concession.

And there'll always be a taker
for a forgotten corner out of **Brueghel**
suitable for a bare-buttocked, incontinent,
sunken-cheeked ending. Little figure
settled in there.

The Sign

First I saw it in colour, then I killed it.
What was still moving, I froze.
That came away. The colour all went
to somebody else's heaven, may they
live on in blessedness. What
came to my hand was fragile, beautiful
and grey, a photograph twilight;
so little to decay there, yet it would
be going down, slowly, be
going down.

Epitaph: Lorine Niedecker

Certain trees
came separately from the wood

and with no special
thought of returning

Occasional Poem 7.1.72

The poets are dying because they are told to die.
What kind of dirt is that? Whose hand
jiggles the nerve, what programme demands it,
what death-train are we on? Not poetry:
some of us drink,
some take the wrong kind of walk
or get picked up in canteens
by killer lays — it's all
tasteless to talk about.
Taste is what death has for the talented. Then
the civilization is filth, its taste
the scum on filth. Then the poets
are going to be moving on out past talent,
out past taste. If taste
gets its gift wrappers on death — well —
out past that, too. There are courts
where nobody ought to testify.

Commuter

Shallow, dangerous, but without sensation:
sun beats in the rear view mirror
with cars squatting in the glare
and coming on. This continues.
Gasholders flicker along the horizon.

Out in all weathers on the test rig
that simulates distance by substituting
a noise drawn between two points;
shallow, my face printed on the windscreen,
profile on the side glass; shallow —

Either I have no secrets
or the whole thing's a secret
I've forgotten to tell myself:
something to make time for on the night run south,
when the dazzle turns to clear black
and I can stare out over the wheel
straight at Orion, printed on the windscreen.

Also

for Derrick Greaves

also there was another story/   a bird suddenly crossing a frame of sky to
alter/   I had no window, the darkness moulded me/   it said the messages
were settled/   we must be crossing a frame of sleep, the sunlit screen over
the matted shadow where the cloud had fallen and gone down lost among
the folds/   and searching for loss more faint than the first loss/   and then
to alter everything by passing it by, asking nothing, expecting nothing to
            alter/   alter/   /also there.

Inscriptions for Bluebeard's Castle

for Ronald King

The Portcullis

Beyond me the common daylight
divides

The Castle

The furthest journey is the journey that stays still
and the light of the sky has come from the world
to be packed for a journey

The Instruments of Torture

Man conceived us, men made us. We work
almost with perfection and we feel no pain.

The Armoury

Provide. That their mouths may bleed into the cinders.
With bronze and steel, provide. With beauty, provide.

The Treasure House

What
the sun touches
shines on forever dead
the dead images of the sun
wonderful

The Garden

Whose is the body you
remember in yourself?

The Land

The light. The rain. The eye. The rainbow —
horizons form, random and inevitable as rainbows
over bright fields of change

The Lake of Tears

Day has turned to a silver mirror
whose dead extent the weeping
eyes could never see

The Last Door

Moonlight the dead image of the day —
here by the light of that last coin
we are alive within an eye:
when the eye closes on us all
it is complete

107 Poems

A scraping in the cokehouse. One red car.
Imperfect science weakens assurances
but swallowing hard brings confidence: fall soft
through to a sunlit verge. Another vision:
stretched out like one expecting autopsy
or showers of sparks across a polished hall.

Swallow all down, to mudstains on the glass;
surmounted by the working, come upon
a sweet for Auntie; for the withdrawn and hurt
something comes sloping upwards, tilts the guard,
then goes across another way: surprise
relaxes from a sideboard in a bottle,
rocks to and fro a while, scores up another —
bottle between the lips — is comforted
into a pointless trip and passes out
finally between two stations, wrapped in yellow.

Sepia slippers in a sepia print,
venerable truth again: it comes direct
and broadens as it comes, is beautiful
if truth is what you want; lies in the blood
and lives on without taint. Magnificent
gorges at sunset! They knew how to live.
They draw us in their footsteps, double-tongued.

To drive under the fog again, and to it,
park by red lights along the road gang's ditch;
changes of *Satin Doll* are getting smothered,
two trumpets and a rhythm section working
carelessly through a roof under the ground;
at twenty past the hour they hit the dirt,
go on across the talk, hit it some more;
a silver surface rears up, wonderful;
somebody scared runs in and turns it over.

Squatting resigned among the rest of it
there's cut and come again; eat anything.
Demolished streets make foregrounds to good skies;
warm hands at rubbish fires, or on a keyboard.
But brightness picks out streaks of signal red,
it's morning. Rumpled, nobody can cope.
What leaks through rotted pipes into the gutter
leaves a long stain that tired arms cannot move,
dispirited by sickness and privation
when peaceful hours have coal dumped under them,
a last delivery, ferried in through sleet.

What's newly made gets treated tenderly;
damage is easy while the aconite
first shows under the window's overhang
and looks well. In the cold light is a refuge,
lying back after breakfast to see birds
flash down the pale grey strip beyond the roof;
and it's a lime-green tent where everything
is fugitive and found, and luminous,
with shadows of a dark track off the calendar
into a depth of sky. Hanging there free,
spiralling down, the ink-trails in the water
that reach the floor and spread. To be well-treated —
a café with net curtains where they bring
coffee or coca-cola to the bedridden —
something to recall on a beleaguered common.
Roads open in succession, windows break;
if both your legs get tired, find a good stick;
slow before lunch, but in the afternoons
Olympic stars perform for invalids
and dark brings in harsh winds and roadside breakdowns;
better to hear of rain on other roofs
or technicolour wrongs worked by hard men.

No choice left but to run, and into it
and back again each time, that being where
the way goes anyhow — so, running
brings it round so much faster, the same dream —
daffodil plastic, various laminates,
children released in yards then sucked away
into an unseen hall; enormous tolerance
somewhere about, and for immediate sky,
hand-lotion-coloured plastic overhead,
the first thing in the world; and back again.
Walking across to the cars in the night air,
everyone slows and vanishes. There'll be
familiar movement when the season dives.
Watch ampelopsis redden the tarred wall;
go straight, and not so fast. The inner sky
is coloured plastic — none the worse for it.

Somewhere the copper pipes a pale gasfitter
left unsecured under the floor tread loose.
The new face might look younger were it not
too harried and too sleepy: there's no time.
Old people go so childish you get scared
thinking about it: someone's moving out.
Under the trees, headlands of alyssum
break through a spring where danger without risk
develops to a style and loses body,
loses its ear for trouble. Ride again.
Desolate sunlit foreshores, visited
and photographed, lie doubly far away;
one more red car gets dealt into the pack;
one guest is laid to rest in his own nature,
his to resist if it should overcome him
travelling in the tracks of a clay lorry
or when the powercut lets the dark back in.

Exhausted, by a different route, twice blessed —
they seem like wooden roses, without yield —
draining the glass again, whatever remains,
past all surprise, repeatedly and strong
though without strength, except to head on out,
surrounded by a street, braced up to feel,
ready for thunder, inescapable change,
the healing of the injured; some idea
of what tradition numbers like these are benched in.

A Grammar for Doctrine

Not what
neutralizes by balance
or by extension cancels —

This is the cleft:
rate it how you will,
as an incredible thing
with tangible properties, even.
But without doubt
the thing that is
shown to people.

Real emphasis
only in the plainness of the signs
for what's known;

mystification itself,
eternal sport,
opens a coloured arch,
plastic to the core,

and a tram with little decoration

sends courageous Gaudi to Avernus.
Entry is quick, you never sense it,

return by way of what's known;
entry is quick, you never sense it,
however you repeat it, or hang by a hair.

Sets

If you take a poem
you must take another
and another
till you have a poet.

And if you take a poet
you'll take another, and so on,
till finally you get
a civilization: or just
the dirtiest brawl you ever saw —
the choice isn't yours.

In the Black Country

Dudley from the Castle keep
looks like a town by Kokoschka,

one town excited
by plural perspectives

into four or five
landscapes of opportunity

each one on offer
under a selection of skies,

and it wheels, dips,
shoulders up, opens away

with clarity and confusion —
Art's marvellous.

Timelessness of Desire

Into the purpose: or out.
There is only, without a tune,
timelessness of desire.

don't open up the way
this town shines in through glass
and the days darken;
there's nothing better,
not one thing better to do —

What's now only disproved
was once imagined.

At Once

I say at once there's a light on the slope among the allotment huts. If I leave
it a moment unsaid it'll set solid, and that only the beginning. But, said, it
has gone.

The wonderful light, clear and pale like a redcurrant, is set off by a
comfortable mist of winter afternoon over to one side of it among the
allotment gardens.

Appearance of mist. The light in its glass. The witnesses were built in about
1910 in the shapes of houses. The stream crawls past the bottom of the
slope, edged with vegetables and crossed by planks. You can approach.

The light is in the earth if anywhere. This is already the place where it was.
We've hardly started, and I want to do it again.

In the Wall

The trails of light all start
from unstable origins
that drift in the dark
in every direction

They coil and wave
into the frame

which is the dark

They make loops of lemon,
of brilliant angelica, streams
of shimmering ruby water

And they stop dead. Arrested,
it turns to a wet street.

Drive at the barrier again. It makes
a night, wet with brilliances.

Brilliant with the power of arrest.

— Or stale, muffled, the senses
having no edge:
feeling for the underside,
wakeful. The name
is Charlatan. A trodden place,
a city: the feet have been

everywhere — on the pillows,
across the benches, on to the walls.

Deep under the viaduct arches
the bare earth is barren;
no rain or daylight. It is dead
dirt. The naked foot,
the soft parts
have been set down here too.

                    Central
to the world, a toilet cubicle
under the street:
                    a judas-hole
in the door, spikes round the top,
a white crisis-chamber.

      In eight or nine distinct
dried brown spattered arcs
somebody's blood has jetted
the whole height of the tiles.
                          He's quite
gone away now.

In the walls

town gods and household gods
used to stare
            out of ringed eyes

seeing
what was never to be said

lacking, in any case,
discourse.

A scent with no face
in darkness:

sallow beyond the skin
and through to a lily-of-the-valley odour
sallow satin

lifted in the voice

not looking but breathing fast

shifting little and quick
more sensible than sense allows;

a raid into the unalterable.

Household god
on a hall table,
stage-lit from a streetlamp
through frosted glass:

eyeless, topless clay head,
true human image
thrown up in Leisure Arts,
holds evenly within his form
a loose mess of papers

Can't speak.
Can't read.

Emblem

for Basil Bunting

Wing
        torn out of stone
like a paper fan

Hung in a sky
          so hard
the stone seems paper

Bare stems of ivy
          silver themselves
into the stones

And hold up the wall
          like an armature
till they force it apart

Corner

Dark projecting corner
of shiny mahogany
standing out
among shadowy walls.

Beside it the face
gleams. Somebody standing

or halted in walking out.
A teacher —
          there are no
teachers here, no lessons.
It's not a teacher.
          Somebody —
the settings are made
to show faces off.
People have to expect to be seen.
They can clear themselves of enigma
if the settings allow,
if the enigma —

Keats's death-mask
a face built out from a corner.

If you're living
any decor
can make a wraith of you.

On the Open Side

On the open side, look out
for sun-patches of sea-blue:

if you see them
it's beginning to shift
with factory towers along the edge,
chalk-white and silver,
empty even of machines

— the other life,
the endless other life,
endless beyond the beginning;

that holds and suddenly presents
a sunny day twenty years ago,
the open window of a train
held up on an embankment for an hour:

down the field there were children playing
round a concrete garage.
That was all. Something the other life wanted —
I hadn't kept it.

                    But look out
for the sea-blue patches.
They'll not make problems.

It is Writing

Because it could do it well
the poem wants to glorify suffering.
I mistrust it.

I mistrust the poem in its hour of success,
a thing capable of being
tempted by ethics into the wonderful.

A Poem Not a Picture

On a ground remarkable for lack of character, sweeps of direction form.

It's not possible to determine whether they rise from the ground's qualities or are marked on to it. Or whether, if the first, the lines suck the ground's force up, or are its delegates; or if the second, whether the imposed marks mobilize or defeat it; or both, in all cases.

Out of a scratch ontology the sweeps of direction form, and, as if having direction, produce, at wide intervals, the events.

These are wiry nodes made of small intersecting planes as if rendered by hatching, and having a vapid, played-out look. But they are the nearest the field has to intense features. Each has a little patch of red.

Cut Worm

You're the invention
I invented once before —
I had forgotten.

                     I need to invent you now
more than you need to be remembered.

Dark on Dark

Dark on dark —
they never merge:

the eye imagines to separate them,
imagines to make them one:

imagines the notion of impossibility
for eyes.

The Only Image

Salts work their way
to the outside of a plant pot
and dry white.

                This encrustation
is the only image.
                The rest —
the entire winter, if there's winter —
comes as a variable that shifts
in any part, or vanishes.

                I can
compare what I like to the salts,
to the pot, if there's a pot,
to the winter if there's a winter.

The salts I can compare
to anything there is.
Anything.

Dusk

The sun sets
in a wall that holds the sky.

You'll not
be here long, maybe.

The window
filled with reflections
turns on its pivot;

beyond its edge
the air goes on cold and deep;
your hand feels it,
or mine, or both;
it's the same air for ever.

Now reach across the dark.

Now touch the mountain.

Mouth-Talk

Mouth of artifice
fashioned to make
mouth-talk.

Not formal. Familiar.

The tablecloth
        falls,
                legs
rest on the floor;
        draperies of the design
everywhere of the most temporary:

it is made.

It has motion:
        the floor a wind
propelled by the thought of a fan—

                (happy the world so made
                as to be
                blessed by the modes of art)

driven by fan blades
and their dark eddies,
        wave-patterns that convert
into mouth-talk.

Wish

That, once sighted,
it should move of its own accord,

right out of view most likely;

and if it does that
a more primitive state altogether
gets revealed:
                    a hardened
paste-patch
of rhubarb and mud-green.

— here comes morning again,
sunshine out of an egg —

and where the first sight
was all design
                    what's left
shows its behind,
stamps and wiggles,
resists transference,
won't be anybody's currency;
doesn't aim to please
and for the most part doesn't.

Without Location

A life without location —
just the two of us
maybe, or a few —

keeping in closeup:
and the colours —
and just the colours

coming from the common source
one after the other
on a pulse;

and passing around us,
turning about and
flaking to form a world,

patterning on the need for a world
made on a pulse.
That way we keep the colours,

till they break and go
and leave no trace; nothing
that could hold an association.

Some Loss

Being drawn again
through the same moment

helpless, and to find
everything simpler yet:

more things I forgot to remember
have gone; maybe because I forgot.

Instead there is blankness
and there is grace:

the insistence of the essential,
the sublime made lyrical
at the loss of what's forgotten.

The Poet's Message

What sort of message —
what sort of man
comes in a message?

I would
get into a message if I could
and come complete
to where I can see
what's across the park:
and leave my own position
empty for you in its frame.

116

Handsworth Liberties

1
Open —
and away

in all directions:
room at last for the sky
and a horizon;

for pale new towers in the north
right on the line.

It all
radiates outwards
in a lightheaded air
without image;

there is a world.
It has been made
out of the tracks of waves
broken against the rim
and coming back awry; at the final
flicker they are old grass and fences.
With special intensity
they gather and break out
through birch-bark knuckles.

2
Lazily into the curve,
two roads of similar importance
but different ages, join,

doubling the daylight
where the traffic doubles,
the spaces
where the new cut through
cleared the old buildings back
remaining clear
even when built on.

3

A thin smoke
in the air as dusk approaches;
unpointed brickwork
lightly soiled,
not new, not old;

papery pink roses
in the smoke.

The place is full of people.
It is thin. They are moving.
The windows
hold up the twilight.
It will be dark, but never deep.

4

Something has to happen here.
There must be change.
It's the place
from which the old world fell away
leaning in its dark hollow.

We can go there
into the seepage,
the cottage garden with hostas
in a chimneypot

or somewhere here
in the crowd of exchanges
we can change.

5

From here to there —
a trip between two locations
ill-conceived, raw, surreal
outgrowths of common sense, almost
merging one into the other

except for the turn
where here and there
change places, the moment
always a surprise:

on an ordinary day a brief
lightness, charm between realities;

on a good day, a break
life can flood in and fill.

6
Tranquility a manner;
peace, a quality.

With not even a whiff of peace
tranquilities ride the dusk
rank upon rank,
the light catching their edges.

Take masonry
and vegetation.
Witness composition
repeatedly.

7
The tall place
the top to it
the arena with a crowd.

They do things by the roadside
they could have done in rooms,
but think this better,

settling amid the traffic
on the central reservation turf,
the heart of everything
between the trees.

And with style: they bring
midnight and its trappings out
into the sun shadows.

8

At the end of the familiar,
throwing away the end
of the first energy, regardless;
nothing for getting home with —

if there's more
it rises from under the first
step into the strange
and under the next and goes on
lifting up all the way;

nothing has a history. The most
gnarled things are all new,

mercurial tongues
dart in at the mouth,
in at the ears;

they lick at the joints. It is new,
this moon-sweat; or by day
this walking through groundsel
among cracked concrete foundations
with devil-dung
in the corners.

Newest of all
the loading platform
of a wrecked dairy,
departure point
for a further journey
into the strangest yet —

Getting home — getting home somehow,
late, late and small.

9

Riding out of the built-up
valley without a view
on to the built-up crest
where a nondescript murky evening
comes into its own

while everybody gets home
and in under the roofs.

A place for the boys,
for the cyclists,
the strong.

10
A mild blight, a sterility,
the comfort of others'
homecoming
by way of the paved strip
down one side of the lane;

the separate streetlamps lead
through to the new houses,
which is a clear way

flanked silently
by a laundry —
brick, laurels, a cokeheap
across from the cemetery gate —
a printing works and a small
cycle factory; hard tennis courts.

The cemetery's a valley
of long grass set with marble,
separate as a sea;

apart from the pavement
asphalt and grit are spread
for floors; there are railings,
tarred. It is all
unfinished and still.

11
Hit the bottom and spread out
among towering structures
and total dirt.

The din compelling
but irrelevant
has the effect of a silence

that drowns out spirit noise
from the sunlit cumulus ranges
over the roofs.

On the way to anywhere
stop off at the old furnace —
maybe for good.

12
Travesties of the world
come out of the fog
and rest at the boundary.

They never come in:
strange vehicles,
forms of outlandish factories
carried by sound through the air,
they stop at the border,
which is no sort of place;
then they go back.

Why do they manifest themselves?
What good does watching for them do?
They come
out of a lesser world.

I shall go with them sometimes
till the journey dissolves under me.

13
Shines coldly away
down into distance
and fades
on the next rise to the mist.

If you live on a slope, the first
fact is that all
falls before anything rises,

and that can be too far away
for what it's worth. I

never went there.

Somebody else did, and
I went with them;
I didn't know why. I remember
coming a long way back
out of the hollow

where there was nothing to see
but immediacy, a long wall.

14
A falling away
            and a rejoicing
      soon after the arrivals —
      small, bright, suspicious —
were complete:
                  strangers
sizing one another up
in front of the shade.

With the falling away
                  the tale finishes.

Before, nobody knew them,
after, there was nothing to know.
They were swept down into the sky
or let to drift along edges
that reached out, finite,
balking the advance, delaying
their disappearance out
into the clear.

15
No dark in the body
deep as this
            even though the sun
hardens the upper world.

                                        A ladder
climbs down under the side
in the shadow of the tank
and crosses tarry pools.
                              There are
metals that burn the air;
a deathly blue stain
in the cinder ballast,
and out there past the shade
sunlit rust hangs on the still water.

Deep as we go
into the stink
this is not the base,
not the ground. This
is the entertainment.

16
This is where the game gets dirty.
It plays
the illusion
of insecurity.

Shops
give way to hoardings,
the ground rumbles,
the street turns to a bridge —
flare and glitter of a roadway
all wheels and feet.

There's no substance;
but inside all this
there's a summer afternoon
shining in a tired room
with a cast-iron radiator,
pipes for a gas fire:
no carpet. No motion.
No security.

Of the Empirical Self and for Me
for M. E.

In my poems there's seldom
any *I* or *you* —

> you know me, Mary;
> you wouldn't expect it of me —

The night here is humid:
there are two of us sitting out
on the bench under the window;

> two invisible ghosts
> lift glasses of white milk
> and drink
>   and the lamplight
stiffens the white fence opposite.

A tall man passes
with what looks like a black dog.
He stares at the milk, and says
> *It's nice to be able*
> *to drink a cup of*
> *coffee outside at night . . .*

and vanishes. So —
What kind of a world? Even
love's not often a poem. The night
has to move quickly. Sudden rain.
Thunder bursts across the mountain;
the village goes dark with blown fuses,
and lightning-strokes repeatedly
bang out their own reality-prints
of the same white houses
staring an instant out of the dark.

Barnardine's Reply

Barnardine, given his life back,
is silent.

With such conditions
what can he say?

The talk
is all about mad arrangements, the owners
counting on their fingers,
calling it discourse, cheating,
so long as the light increases,
the prisms divide and subdivide,
the caverns crystallize out into day.

Barnardine,
whose sole insight into time
is that the right day for being hanged on
doesn't exist,
                is given
the future to understand.

It comes
as a free sample from the patentholders;
it keeps him quiet for a while.

It's not the reprieve in itself
that baffles him:
he smelt that coming
well before justice devised it —
                            lords

who accept the warrant,
put on a clean shirt,
walk to the scaffold,
shake hands all round,
forgive the headsman,

kneel down and say, distinctly, 'Now!' attract
pickpockets of the mind —

But he's led away
not into the black vomit pit
he came out of
but into a dawn world

of images without words
where armed men, shadows in pewter,
ride out of the air and vanish,
and never once stop to say what they mean:

— thumb with a broken nail
starts at the ear lobe,
traces the artery down,
crosses the clavicle, circles
the veined breast with its risen nipple,
goes down under the slope of the belly,
stretching the skin after it —

  butchered just for his stink,
  and for the look in his eye —

In the grey light of a deserted barn
the Venus, bending to grip the stone sill,
puts up no case for what she's after,
not even a sigh,
but flexes her back.
     No choice for the Adonis
but to mount her wordlessly, like a hunting dog —
  just for her scent
  and for the look in her eye.

Somebody draws
a Justice
on the jail wall;

gagged with its blindfold
and wild about the eyes.

Simple Location

In simple location
the sticks take fire:

they cross and tangle
with smoke-spurts

breaking into the sunlight
as it strikes the ground,

coming in from a fog-rim
through the bleached grasses;

and if a golden drop escapes
anywhere on the skin

of a boy I've seen starting to sweat
in my dream

it has its place —
or if it should leave his eye

by way of the honey-crust there
and slowly trickle

down by the corner of his mouth
to undo everything:

if the sense of charged confinement
should come again

it seizes on a breath
caught in its place in the body,

held there a moment; still
filled with the fire-scents.

If I Didn't

If I didn't dislike
mentioning works of art

I could say
the poem has always
already started, the parapet
snaking away, its grey line guarding
the football field and the sea

— the parapet
has always already started
snaking away, its grey line
guarding the football field and the sea

and under whatever progression
takes things forward

there's always
the looking down
between the moving frames

into those other movements
made long ago or in some
irrecoverable scale
but in the same alignment
and close to recall.

Some I don't recognize,
but I believe them —

one system of crimson scaffolding,
another, of flanges —

All of them must be mine,
the way I move on:

and there I am,
half my lifetime back,
on Goodrington sands
one winter Saturday,

troubled in mind: troubled
only by Goodrington beach
under the gloom, the look of it
against its hinterland

and to be walking
acres of sandy wrack,
sodden and unstable
from one end to the other.

Paraphrases

for Peter Ryan

Dear Mr Fisher I am writing
a thesis on your work.
But am unable to obtain
texts. I have articles by Davie, D.,
and Mottram, E.,
but not your Books since booksellers
I have approached refuse to
take my order saying they
can no longer afford to
handle 'this type of business'. It is
too late! for me to change
my subject to the work of a more
popular writer, so please Mr Fisher
you must help me since I face the alternatives
of failing my degree or repaying
the whole of my scholarship money . . .

Dear Mr Fisher although I have been unable
to read much of your work (to get it that is)
I am a great admirer of it and your landscapes
have become so real to me I am convinced I have, in fact,
become you. I have never, however,
seen any photograph of you, and am most curious
to have an idea of your appearance,
beyond what my mirror, of course, tells me.
The cover of your *Collected Poems*
(reproduced in the *Guardian*, November 1971)
shows upwards of fifty faces; but which is yours? Are you
the little boy at the front, and if so have you
changed much since then?

Dear Mr Fisher recently while studying
selections from a modern anthology with
one of my GCE groups I came across your interestingly titled
'Starting to Make a Tree'. After the discussion I felt strongly
you were definitely *holding something back* in this poem
though I can't quite reach it. Are you often in Rugby?
If you are, perhaps we could meet and I could
try at least to explain. Cordially, Avis Tree. PS. Should we
arrange a rendezvous I'm afraid I wouldn't
know who to look out for as I've never unfortunately

seen your photograph. But I notice you were born in 1930
the same year as Ted Hughes. Would I be right
in expecting you to resemble *him*, more or less?

                     — Dear Ms Tree,
It's true I'm in Rugby quite often, but the train
goes through without stopping. Could you fancy standing
outside the UP Refreshment Room a few times so that
I could learn to recognize *you*? If you could
just get hold of my four books, and wave them,
then I'd know it was you. As for my own appearance
I suppose it inclines more to the
Philip Larkin side of Ted Hughes's looks . . .
See if you think so as I go by . . .

Dear Mr Fisher I have been commissioned
to write a short
critical book on your work
but find that although I have a full
dossier of reviews etcetera
I don't have access to your books. Libraries
over here seem just not to have bought them in.
Since the books are quite a few years old now
I imagine they'll all have been remaindered
some while back? Or worse, pulped? So can
you advise me on locating second-hand copies,
not too expensively I hope? Anyway,
yours, with apologies and respect . . .

Dear Mr Fisher I am now
so certain I am you that it is obvious to me
that the collection of poems I am currently working on
              must be
your own next book! Can you let me know —
who is to publish it and exactly when
it will be appearing? I shouldn't like there to
be any trouble over contracts, 'plagiarism'
etcetera; besides which it would be a pity
to think one of us was wasting time and effort.
How far have *you* got? Please help me. I
do think this is urgent . . .

Passing Newbridge-on-Wye

All the space under the bridge
fills with the light
of the bare ash-trees and the stone:

what glitter the softness has
comes from the February sun
striking across the pebbles of the riverbed;
there's nothing else.

   The pale light bursts
the distance of the valley as if
in a water-drop on a windscreen,
but in a fullness
with no sharp instant of design:
it's not for catching.

              The turn
towards the south at the bridgehead
dissolves to a state of the air,

a state the road rests into
as it passes over.

So you can be clouded
with clarities in the act
of crossing the undemanding water.

Diversions

1
Trouble coming, on a Saturday or a Monday,
some day with a name to it:

staining the old paths trouble knows,
though I forget them.

2
Walk through, minding the nettles
at the corner of the brick path —

don't feel sorry for language, it doesn't bear
        talking about.

3
Built for quoting in a tight corner —
The power of dead imaginings to return.

4
Just beside my track through the dark,
my own dark, not to be described,
the screech-owl
sounds, in his proper cry
and in all his veritable image —
you would know him at once.

Beyond him
a dissolution of my darkness
into such forms
as live there in the space
beyond the clear image of an owl:

forms without image;
pointless to describe.

5
I saw what there was to write and I wrote it.
When it felt what I was doing, it lay down and died under
        me.

6
Grey weather beating across the upland,
and the weather matters.
Grey weather beating easily across the upland.

7
Crooked-angle wings
blown sideways
against the edge of the picture.

8
Roused from a double
depth of sleep, looking up
through a hole in the sleep's surface above;
no sense of what's there;
a luminous dial
weaves along the dark like a torch.
There's somebody already
up and about, a touch-paper crackle
to their whispering

9
The kites are the best sort of gods,
mindless, but all style;

even their capriciousness,
however dominant,
not theirs at all.

Lost from its line
one flies steadily out to sea,
its printed imperturbable face
glinting as it dips and rises
dwindling over the waves.

The crowd on the shore
reach out their hearts.

10
Leaden August with the life gone out of it,
not enough motion
to shift old used-up things.

A bad time to be rid of troubles,
they roll back in.

Dead troubles take longer than live ones.

11
The pilgrim disposition —
walking in strung-out crowds
on exposed trackways
as if ten yards from home:

domestic to-ing and fro-ing
uncoiled and elongated
in a dream of purpose.

12
Then some calm and formal portrait
to turn a level gaze
on the milling notions,
its tawniness of skin denoting
tension maybe, a controlled pallor;

or a blush of self-delight
welling softly from its intelligence.

13
Periodicity: the crack
under the door of this room
as I stare at it, late at night,
has the same relation to its field
as — what?

The corner over the curtain-rail
in a room I was in one night
forty years ago and more.

The light and the height are different,
and so am I;
but something in the staring
comes round again.

              So I stare
at the single recurrence of a counter
I expect never to need.

14

Sliding the tongue-leaved
crassula arborescens
smartly in its pot and saucer
from one end of the windowsill
right down to the other

alters the framed view, much
as a louvred shutter would.

All my life I've been left-handed.

15

Here comes the modulation.
Elbows in, tighten up:
a sucked-in, menacing sound,
but full. The space is narrow,
the time marked out,
and everybody's watching.

16

The woman across the lane
stoops, hands on knees,
behind out, black and grey hair
falling forward, her nose level
with the top of a four foot wall
under a huge shaggy bank of privet.

Nose to nose with her across a saucer,
his tail lifting into the privet shadow,
a big dark cat with a man's
face marked out in white.

Quietly,
in a good, firm Scottish voice,
set well down,
she tells his story:
                        how
when his owners first
moved another cat in on him
then moved out altogether,
he ran wild for three years,

haunting back once in a while,
a frightener.

After that
for a year and a half
she'd set for him daily,
slowly drawing him in

as near as this;
she didn't expect more.

She talks, and the cat drinks.
He turns his mask to me,
sees me, and without pausing
vanishes.

      Later, from a distance,
I see the two of them again,
a saucer apart. The cat
with his enormous guilt
and importance;
        the fortunate cat,
to have such a calm Scots lady
to understand his importance.

17
Out to one side
a flight of shops
turning towards the sun,

each one a shallow step higher,
white and new and good.

And there's the ultimate in shops;
the gallery.

Somebody can be stood —
can elect to stand —
in fresh clothes but barefoot
on a slate ledge, in the place of a pot,

fastidious
beyond the flakings of the skin,

the vegetable variants of body-form,
the negative
body-aura,
that shadowy khaki coat.

Can stand, and receive attributions
of pain and excellence.

18
Everything cast in iron
must first be made in wood —

The foundry patternmaker
shapes drains, gears,
furnace doors, couplings
in yellow pine.

His work fulfils the conditions for myth:
it celebrates origin,
it fixes forms for endless recurrence;
it relates energy to form;
is useless in itself;

for all these reasons it also attracts
aesthetic responses in anybody
free to respond aesthetically;

and it can be thought with;

arranged on trays in the Industrial Museum,
it mimes the comportment
of the gods in the Ethnology cases.

19
Outlines
start to appear
on the milky surface.

Points first,
quickening into perimeters
branches and dividers;
an accelerating wonder.

Arrest; try lifting it away
before the creation
diversifies totally
to a deadlocked fission:
diamond-faceted housebricks
in less than light.

The thin trace lifted off will drop
into a new medium and dissolve.
On the bland surface
will appear new outlines.

Both these ways are in nature.

20
A world
arranged in zones
outside and into
this waterfront café.

A strip of sky
misty with light,

a deep band of
dark hazy mountainside,

a whole estuary width
foreshortened almost to nothing,

a quay,
a full harbour;

then a pavement,
a sill,

the table where I sit,
and the darkness in my head.

Everything still along its level

except the middle zone, the harbour water,
turbulent with the sunlight
even in calm air.

Rules and Ranges for Ian Tyson

Horizons release skies.

A huge wall has a man's shoulder. In the only representation we have it is mottled with a rash and distorted overall, seen through gelatine.

The Thames with its waterfronts; a fabric with a Japanese Anemone design. They intersect at Chaos.

The force of darkness is hard, rigid, incapable of motion either within its own form or by way of evasion. All the same, it is very difficult to find.

The experience of a wind, as if it were a photogravure made of dots. To be vastly magnified.

To walk along two adjacent sides of a building at once, as of right.

After a fair number of years the distasteful aspects of the whole business become inescapable. Our frustrations will die with us, their particular qualities unsuspected. Or we can make the concrete we're staring at start talking back.

Watch the intelligence as it swallows appearances. Half the left side, a set of tones, a dimension or two. Never the whole thing at once. But we shouldn't need to comfort ourselves with thoughts like this.

Under the new system some bricks will still be made without radio receivers or photo-electric elements. No potential for colour-change, light-emission, variation of density — just pure, solid bricks. They'll be special.

A terror ruffling the grass far off, and passing without coming near. Between that place and this the grass is a continuous stretch with no intervening features.

Under the new football rules the goals will be set, not facing each other down a rectangle but at the centres of adjoining sides of a square pitch, and the teams will be arranged for attack and defence accordingly. Some minor changes in rules are bound to be necessary, but there will also be rich variations in styles of play.

Darkness fell, surrounding and separating the hollow breves. They howled and shone all night.

Staffordshire Red

for Geoffrey Hill

There are still clefts cut in the earth
to receive us living:

the turn in the road, sheer through
the sandstone at Offley
caught me unawares,
and drew me, car and all,
down in the rock

closed overhead with trees
that arched from the walls,
their watery green
lighting ferns and moss-shags.

I had not been looking for the passage,
only for the way;

but being suddenly in
was drawn through slowly

— altering by an age,
altering again —

and then the road dropped me
out into a small, well-wooded
valley in vacancy.
Behind me.
was a nondescript cleft in the trees.
It was still the same sunless afternoon,
no north or south anywhere in the sky.
By side roads
I made my way out and round again
across the mildnesses of Staffordshire
where the world changes with every mile
and never says so.

When I came face to face with the entry
I passed myself through it a second time,
to see how it was.

It was as it had been.

The savage cut in the red ridge,
the turn in the traveller's bowels,
by design ancient or not;
the brush-flick of energy
between earth and belly;
the evenness of it. How hard
is understanding? Some things
are lying in wait in the world,
walking about in the world,
happening when touched, as they must.

The Dirty Dozen

Dirty Nature,
Dirty sea;

Dirty daytime,
Dirty cry;

Dirty melody,
Dirty heart;

Dirty God,
Dirty surprise;

Dirty drumlin,
Dirty design;

Dirty radiance,
Dirty ghost.

Style

for Michael Hamburger

Style? I couldn't begin.
That marriage (like a supple glove
that won't suffer me to breathe)
to the language of one's time
and class. The languages
of my times and classes.

Those intricacies
of self and sign. The power to mimic
and be myself. I couldn't.

I'd rather reach the air
as a version by my friend Michael.
He knows good Englishes.
And he knows the language
language gets my poems out of.

Butterton Ford

— But the street is the river —

Not much of a river, and
not much of a street.
Not much call for either
down this end.

They may as well double up,
let the stream slant out of the hill
over the cobbles in a thin race;

a footway for flood-time,
with a little bridge with overflow holes
cut in the coping. Keep the pressure off.

3rd November 1976

Maybe twenty of us in the late afternoon
are still in discussion. We're talking
about the Arts Council of Great Britain
and its beliefs about itself. We're baffled.

We're in a hired pale clubroom
high over the County Cricket Ground
and we're a set of darkening heads,
turning and talking and hanging down;

beyond the plate glass, in another system, silent,
the green pitch rears up, all colour,
and differently processed. Across it in olive overalls
three performance artists persistently move
with rakes and rods. The cold sky steepens.
Twilight catches the flats rising out of the trees.

One of our number is abducted
into the picture. A sculptor innocent of bureaucracy
raises his fine head to speak out;
and the window and its world frame him.
He is made clear.

Discovering the Form

Discovering the form of vibrancy
in one of the minor hilltops,

the whorl of an ear
twisting somewhere under the turf,
a curve you have to guess at.

In a house out of sight round the shoulder,
out of ordinary earshot,
a desperate mother, shut in with her child,
raves back at it when it cries,
on and on and on, in misery and fear.

Round on the quiet side of the hill
their shrieks fill an empty meadow.

The Trace

Although at first it was single
and silver

it travelled as ink falls
through cold water

and gleamed in a vein
out of a darkness

that turned suddenly on its back
and was dusty instead

letting go forth as it must
a plummet of red wax

from whose course when they lost it
rings of dull steel

like snake ribs in a sidelong curve
twisted away and lifted

to clamp on to a concrete
precipice broken with rust

and with shrubby growths
clustering under it

their leaves
shading and silvering

in the currents of light
draining among the branches

to where it was sodden
full of silky swallowed hair

that dried and was
flying in a fan

air flickering from its ends
collecting silvers

as it twined itself
into the gauze

then scattered as many
mercurial bolts

all through the chamber
darting everywhere

in under the roof-keel
with its infinite brown decline

its warp
unmistakable as it reached

down into the daylight
with a sidelong wooden nose

biased like the set
of a rudder and pulling in

everything that could raise
a bright wave against it

making colours print
themselves on to planes with the effort

one plate of red enamel
dominant and persisting

even through a grey
sleet that scored its face

running away as water
welling downwards

through a raised irregular
static vein

moving only by rills
within itself.

Five Pilgrims in the Prologue to the Canterbury Tales

Knight

He bore himself, or the self he had was borne,
through great indignities and darkness
inviolate in a glass braced with silver.
It was as small as that.

In Lithuania, spars and foul fabrics
prised out of black silt with a frozen crack;

elsewhere, excrements rolled in sand;

scattered all across the back of beyond, seedy nobility,
mincing and cheating, and getting cut to pieces.

As often as he set forth, he would find himself
returning through strange farmlands
in incomprehensible weather.

Webbe

Risen by weaving works to this
abstraction: something at the same time
rich and dry about that trade,
a station on the road to pure money;
stretching resources till the patterns glow.

Dyere

Even in the pools
and vats, even in the steamy
swags and folds and leaf-stinks
stuck in the nose at night
there's always been someone's white finger
waving above it all.

Tapicer

Base yourself on these,
Lords. I have slipped
the carpet in under your foot,
the cushion beneath your buttocks;
the walls are all taken care of.

It is with this sort of confidence
you can look me in the eye.

Cook

Slow cooking in a world on the move.
The hours are slow, the pies live a long time;
cooling and warming, they ride the bacterial dumb-waiter.
A cook goes in and out of focus many times as he learns.

On the bright side, the ulcer's almost forgettable,
and the flies loose in the shop don't irritate,
being mostly middle-sized and black, and in no hurry.

Releases

One Sunday afternoon by the shuttered corner shop heavy flies, bright green and blue, swarmed on the fish boxes. Culling them with a rolled paper you lived close to them. Dead, they filled up the Maltese-cross grooves in the slate-blue paving bricks.

The gantries of the travelling cranes hit back at the sky in the afterglow.

All structures are mysterious, however the explanation goes.

In a place like that, virtually everything is a structure.

Wherever the floor or the crust of the ground is opened, indoors or out, there is revealed some part of the continuous underwork: a tarry pipe, a gas main, a pot of still yellow water.

Grammar of my journey through the streets, through the rooms and halls. I sense my movement always as forward and express it so. I ignore my ability to turn left or right. Or to glance, or to think, aside. My whole tortuous track warps into a single advancing line. This warping hardly shows in the stations of my path, even the unspiral staircase, but it has many consequences in what lies beyond or beside what I travel through. All the same, I'm prevented by definition from knowing what those consequences are like. If I even think about them they swing into place directly ahead of me, and in doing so they straighten themselves up.

The greater part of my life is past, and I seem to have done nothing. Yet I've achieved rather more than I've attempted, so that means I've kept my standards.

It's amazing what you can say if you try.

Forgive me, but on the hillsides over Abergavenny the houses are too many, too high and too white.

Far into the jigsaw puzzle there are blobs and flecks of whitish grey in the dark inks of the hillside. The puzzle shows anarchy and nature to the child. It's not for solving, it's for the disordering of feeling. Some piece of going-on substance, sky-stuff, skyline-stuff, some locus with no qualities of its own, is the heart of the world's energies. Some reason, at the mercy of the laws of a puzzle, assigns those energies. The puzzle helps teach a fascinated resignation: any pavement, any dwarf wall, any old inside leg. The best isn't necessarily best.

The double-fronted shop perched next to the road bridge over the main line cutting, its shutters peeling down to the greeny-blue undercoat in the sun. Shadowy, with an old Kelvinator ice-cream machine and many groceries, but with plenty of space above them. Not a popular shop. Nobody could ever lay a story on it. The story would recede, disperse, fail to answer when called.

Coming down the hill by the park early one summer evening on the top deck of a bus, seeing the warm roofs, the horse-chestnuts and limes, stretching out from me, spreading under me; deciding this was the first thing I needed, and the strongest. What I thought was not what I knew was there, but I stayed.

On leaving, photographing a few houses, street corners, the railway bridge, on a bright afternoon, in case they're about to go. Colours welling suddenly through all the shapes that had never seemed coloured at all.

The Red and the Black

Most of the houses were brick
and soft by night. But these were all done

with timber painted
gloss black across white plaster. Big, they were

stepping down
both sides of the steep street

towards the skyline out there
and the moon coming up. Yes,

there were crimson curtains all right,
and cushions of the same. After

the war. Coming up to the next war.
Between the wars. Between times.

The park lake in the suburb has been
honoured up into art, flat

on the back of a playing card.
The poplars in silhouette are climbing

nearly to the gold moon that floats
in scarlet. The houses,

dense black, with gold and scarlet windows,
have reared high into the poplars;

a thin rendering of *Clair de Lune* gets
drawn acrosss the night as it is condemned to be.

Don't think I'm being patronizing.
The women's wiry hairs

are real enough, moist in the dark,
shifting on painted benches

among the graves, lipping and tugging
at the fingers' ends. The day

you're patronizing to a woman's cunt
you're faded, you're a kidney

dropped in the gutter, you're a dried
bag. That's the faith of it, at least.

A faith against what odds? Deciding
she needed glasses she'd appear

glimmering with finish, staring
brightly through the black frames,

hair cut to a knife of ink,
neat lips of crimson; but bringing

the same good-natured
perfumed crotch, that was

tangy, commercial, ambitious,
epithet-hungry, effective, timeless,

whatever Lawrentian strictures
came bubbling from her head.

Albion the cinema had a garden
where the orchestra used to be.

And it was always there in the dark
under the guns, or the light

from the transparent ruched curtain
that held floods of colour

or from huge white faces
flapping over its canvas arches.

It had low walls, and something
like a fountain,

and it was modest,
with hand-built hollyhocks.

Between shows they used to play a blue
moonlight over its pastel stones.

A hill of galantine
waits for me in its frame.

In a black and gilt photo frame
it's with me always, in colour:

steep, a flat-topped tumulus,
God's Grave,

a mound that might sit with authority
on a low hill among tall hills

green under a rolling sky
and seeming to offer the promise

of being put to the breast
every day of a long life.

If it were earth, that's how it would be.
But this is meat

shining in parsley-green and white
and part sliced away.

Solid right through, rich pink
shading to red, cream-veined,

filling the black rectangle
behind the glass, it feels

dense with its energies,
and the fork

stuck in its crown
must never once have quivered.

The green and white,
the meat, the frame,

vehement, inert,
endure. I was meant to

glance at it all just
once and move on

but I wouldn't. I fixed it.
It tried to fix me. The mound

goes rising into the sky backwards,
bitter in its green crust;

into the stormy sky of woodstain
shot with petal flares

and as it lifts obscures the mimic
shape of its own cut section

rises oblique across it and faster,
satellite, or utter independent,

marbled with rose —
pink and terra-cotta gases,

advancing over or voyaging into
that which is not itself,

limpid element,
shattered wherever the bow wave,

ghosting forward, strikes it,
into a sudden precipitate

of opposite green particles
heaped so that it has to

push through and leave a widening
tail as the event passes over.

Wonders of Obligation

We know that hereabouts
comes into being
the malted-milk brickwork
on its journey past the sun.

The face of its designer
sleeps into a tussocky
field with celandines

and the afternoon
comes on steely and still
under the heat,

with part of the skyline
settling to a dark slate
frieze of chimneys
stiffened to peel away
off the western edge.

I saw
the mass graves dug
the size of workhouse wards
into the clay

ready for most of the people
the air-raids were going to kill:

stil at work, still in the fish-queue;
some will have looked down
into their own graves on Sundays

provided
for the poor of Birmingham
the people of Birmingham,
the working people of Birmingham,
the allotment holders and Mother, of Birmingham.
The poor.

Once the bombs got you
you were a pauper:
clay, faeces, no teeth; on a level
with gas mains,
even more at a loss than before,
down in the terraces between the targets,
between the wagon works
and the moonlight on the canal.

A little old woman
with a pink nose, we knew her,
had to go into the pit, dead of pneumonia,
had to go to the pit with the rest,
it was thought shame.

Suddenly to go
to the school jakes with the rest
in a rush by the clock.
What had been strange and inward
become nothing, a piss-pallor
with gabble. Already they were lost,
taught unguessed silliness,
to squirt and squeal there.
What was wrong? Suddenly
to distrust your own class
and be demoralised
as any public-school boy.

The things we make up out of language
turn into common property.
To feel responsible
I put my poor footprint back in.

I preserve
Saturday's afterglow
arched over the skyline road
out of Scot Hay:

156

the hare
zig-zagging slowly

like the shadow of a hare
away up the field-path

to where the blue
translucent sky-glass
reared from the upland
and back overhead

paling, paling
to the west
and down to the muffled rim of the plain.

As many skies as you can look at
stretched in a second
the manifest
of more forms than anyone could see

and it alters
every second you watch it,
bulking and smearing the inks
around landlocked light-harbours.

Right overhead, crane back,
blurred grey tufts of cloud
dyeing themselves blue,
never to be in focus, the glass
marred. Choose this sky. It is
a chosen sky.

What lies
in the mound at Cascob?
The church built into the mound.

In the bell-tower
is in the mound.

Stand
in the cold earth with the tower around you
and spy out to the sanctuary
down to whatever lies dead there

under the tiny crimson
lamp of the live corpse of the god.

Later than all that
or at some other great remove
an old gentleman
takes his ease on a shooting stick
by the playground on Wolstanton Marsh.

A sunny afternoon on the grass
and his cheeks are pink,
his teeth are made for a grin; happily
his arms wave free. The two stiff
women he has with him in trousers and anoraks
indicate him. They point
or incline towards him. One
moves a good way along the path, stretching a pattern.
The cars pass
within a yard of him. Even so,
he seems, on his invisible stick, to be sitting
on the far edge of the opposite pavement.
Numerous people
group and regroup as if coldly
on a coarse sheet of green.

Parked here, talking,
I'm pleasurably watchful
of the long
forces angled in.

The first farmyard I ever saw
was mostly midden
a collapse of black
with dung and straw swirls
where the drays swung
past the sagging barn.
Always silent. The house
averted, a poor ailanthus
by its high garden gate and
the lane along the hilltop
a tangle of watery ruts
that shone between holly hedges.
Through the gaps you could see
the ricks glowing yellow.

The other farm I had
was in an old picture book,

deep-tinted idyll with steam
threshers, laughing men,
Bruno the hound with his black muzzle,
and the World's Tabbiest Cat.

Describing Lloyd's farm now
moralizes it; as the other
always was. But I swear
I saw them both then
in all their properties,
and to me, the difference was neutral.

As if from a chimney
the laws of the sky go floating
slowly above the trees.

And now the single creature
makes itself seen,
isolate,
is an apparition

Near Hartington
in a limestone defile
the barn owl
flaps from an ash
away through the mournful afternoon
misjudging its moment
its omen undelivered.

The hare
dodging towards the skyline at sunset
with a strange goodwill —
he'll do for you and me.

And *mormo maura*
the huge fusty Old Lady moth
rocking its way up
the outside of the dark pane
brandishing all its legs, its
antennae, whirring wings,
zig-zagging upwards, impelled
to be seen coming in from the night.

Now I have come
through obduracy
discomfort and trouble
to recognize it

                    my life keeps
leaking out of my poetry to me
in all directions. It's untidy
ragged and bright
and it's not
used to things

mormo maura
asleep in the curtain
by day.

Scent on the body
inherent or applied
concentrates the mind
holds it from sidelong wandering.
Even when it repels
it pushes directly.

Streaks of life
awkward
showing among straw tussocks
in shallow flood.

Neither living nor saying
has ceremony or bound.

Now I have come
to recognize it, the alder
concentrates my mind
to the water
under its firm green.

Fetching up with
leaf-gloss against
the river-shine.

I want
to remark formally, indeed

stiffly, though not complaining,
that the place where I was raised
had no longer deference for water
and little of it showing. The Rea,
the city's first river,
meagre and under the streets;
and the Tame
wandering waste grounds,
always behind
some factory or fence.

Warstone Pool in the fields
I realized today was a stream dammed
to make way for the colliery.
Handsworth Park lake, again a dam
on the Saxon's
nameless trickle of a stream
under the church bluff. The brook
nearest home, no more than a mile,
ran straight out into the light
from under the cemetery;
and there the caddis-flies would case
themselves in wondrous grit.

I'm obsessed
with cambered tarmacs, concretes,
the washings of rain.

That there can come a sound
as cold as this across the world
on a black summer night,

the moths out there impermeable,
hooded in their crevices
covered in the sound of the rain
breaking from the eaves-gutters
choked with pine needles;
the slippery needles wash everywhere,
they block the down-spouts;
in the shallow pool on the porch roof,
arranged among dashed pine branches
and trails of needles,
I found two ringdove squabs

drowned and picked clean,
dried to black fins.

Fine edge
or deflection
of my feeling towards

anything that behaves or changes,
however slowly; like
my Bryophyllum *Good Luck*,
raised by me from a life-scrap and
now lurching static from its pot,
its leaves winged
with the mouse-ears of its young.
I'm vehemently and steadily
part of its life.

                       Or it slides
sideways and down, under my suspicion —
Now what's it doing?

Suddenly to distrust
the others' mode;
the others. Poinsettias or moths,
or Kenny and Leslie and Leonard,
Edie and Bernard and Dorothy,
the intake of '35; the story of the Wigan pisspot
of about that time, and even
Coleridge's of long before:

I have to set him
to fill it by candlelight
before he transfigures it;

with *mormo maura* the Old Lady moth
beating on the pane to come in.

The Supposed Dancer

Jumping out of the straw,
jumping out of the straw,
his cheeks alive with bristles,
jumping out —
                you've got him!
He'll ruin the lot of you,
jumping out:
      two drums
      tied up in this
jumping
      and the joy and the
jumping
      bitchery of art, the men
      with warm hard
      hearts
              out of the straw —
Got him!
         Alive with bristles,
jumping with his cheeks
out of the straw alive
with bristles, his cheeks
      ready to jump
the cameraman's guitar, the guitarist's
      camera;
          alive with
jumping out of the
jumping out of the straw.

Pete Brown's Old Eggs Still Hatch

Birmingham, October '80. A man
with a pink unwrapped artificial
arm under his arm.

limps into Ladbroke's.

Rudiments

One half-dark night my father
hoisted me, a three-
year-old illiterate,
on to a parapet to look at
a thing he called *The Barge* —
a pun of shapes from *The Prelude* as it turns out,
deranged with long transmission
and still a few years early for encoding
in what were to be Klee's last paintings:

projecting motionless
down there between the bridge supports
there was a dead black *V* in the murk,
gapped, with its bad face upturned.
Behind us, the biggest thing I'd ever seen,
the dark gas-holder
filled up the sky.

One hot afternoon he led me
to a broken plank in a hoarding, then
showed me, pointing straight down,
The Canal itself. A black
rippled solid, made of something
unknown, and having the terrible property
of seeming about to move,
far under our feet. I'd never
seen so much water before.

The Open Poem and the Closed Poem

1

To come out ready capitalized, with outlines,
cross-beams and a display,
and this terrible year
moralizing itself at my feet, right
from the frozen clutter at its beginning
through to this hundred–
and–ninth day since I last
opened this writing notebook
or thought about what was in it;
having written *Roller* and lost it
and forgotten it, and *Wonders*
of Obligation and forgotten every word of it
until asked, and *The Red and the Black*
and forgotten even the title, and *Releases*
and forgotten I'd ever written it
or what it was; and climbed the mysterious
pip of a hill over Oakenclough
in the snow and mud
and wrecked my shoes, and
forgotten where I'd been to wreck them
until reminded, and having run
my new car twelve thousand miles without
memorizing the tyre pressures,
and having lost my room for manoeuvre
to the monetarists, and having
walked Avebury and forgotten all
but the shape of my understanding of it;
to have things clear, the circumstances
answerable for a start, so that it's plain who's talking.

2

Winking drop on the lens
shatters a soft
fog of lamplight in the dark
that hides how close
overhead the wet balks are;
where you're standing, the way
you're standing
makes the signal, what gives

the wires to shine, a handsbreadth
at a time, rapidly sliding
through the thick of the wall, the bank,
black peat that holds
flakes of thrush-egg. Underfoot
in the shallows every unevenness
crawling with rust, alive
with rust. Foot of the signal.

Irreversible

The *Atlantic Review* misspelled Kokoschka.
In three weeks he was dead.

Ninety-three years to build a name —
Kokoschka — but he felt
that fine crack in the glaze.

Then he 'suffered a short illness';
that's what the illness was.
Irreversible.

John Ashbery should watch out.
Hiding as John Ash in Haight-Ashbury
won't help in the clash
of Haight-Asch with John Ashbury;
it's got to happen.

I'm just the maker
of mutant poems. In one
sails became *snails* — try it. With me
Organic Form overproduces. Here's
my poem *The Trace*, that's started
to feed off itself, and breed:

— *silky swallowed hair*
that dried and was
flying in a fan —

Now it's *dined and was*
flying in a fan —

— sulky squalid whore
that dined and was
frying in a fin —

For *Trace* read *Truce;*
for *Bruce, Brace:*

— crying in a fit —

Chisellers! cut deep
into the firm, glistening
sand —

Norseman, pass by!

The Home Pianist's Companion

Clanging along in A-flat
correcting faults,

minding the fifths
and fourths in both hands
and for once
letting the tenths look out
for their own chances,

thinking of Mary Lou,
a lesson to us all,

how she will trench and
trench into the firmness of the music
modestly;

thinking,
in my disorder of twofold sense,
or finding rather
an order thinking for me as I play,
of the look of lean-spoked
railway wagon wheels
clanging on a girder bridge,
chopping the daylight, black
wheel across wheel, spoke

over rim, in behind girder and out
revealing the light, withholding it,
inexorable flickers
of segments in overlap
moving in mean elongate
proportions, the consecutive
fourths of appearances,
harsh gaps, small strong
leverages, never still.

The sour face
on that kind of wheel:
I've known that
ever since I first knew anything;
a primary fact of feeling,
of knowing how
best to look after yourself.

Clanging along in A-flat, and
here they come,
the apports, the arrivals:
fourths, wheel-spokes,
and rapidly the eternal
mask of a narrow-faced cat,
its cornered, cringing intensity
moving me to distraction again.

But into the calm
of a time just after infancy
when most things were still
acceptable

this backward image-trail
projects further
on a straight alignment
across what looked to be emptiness,
checked as void

and suddenly locates the dead,
the utterly forgotten:

primal figure of the line,
primitively remembered,
just a posture of her, an apron,
a gait. Vestigial figure,
neighbouring old woman,
gaunt, narrow-faced, closed in,
acceptable,
soon dead.

Still in the air
haunting the fourths
of A-flat major
with wheels and a glinting cat-face;

reminding me
what it was like to be sure,
before language ever
taught me they were different,
of how some things were the same.

The Whale Knot

Sea-beast for sky-worshippers, the whale
easily absorbs all others.
Colours, languages, creatures, forms. Read
the whale in all the ways clouds
are read. The clouds out of sight
are patterned and inscrutable; chaos
from simple constituents,
form out of simple chaos.

A long-drawn complicity with us all
in the sperm-whale's little eye;
among its cells, somewhere,
land-knowledge, the diverse, our condition.

Decamped into boundless viscosity,
our Absolute,
the whale seems simpler than it is:
as easy water-to-land knot
in the museum sperm-whale's bared
head-bone, alive

as the megaliths are alive, all
the force-lines crossing
within their singular undemanding
forms. Lifted from the whale-head,
a disused quarry
swims, borne on the earth;
its cliffs a moon-cradle,
its waters part of the sky.

The Burning Graves at Netherton

This is a hill that holds the church up.

This is a hill that burned
part of itself away:

down in the coal measures
a slow smoulder

breaking out idly at last
high on the slope

in patches
among the churchyard avenues.

Netherton church lifts up
out of the falling land below Dudley;

on its clean promontory
you see it from far off.

Not burning. The fire
never raged, nor did the graves flame
even by night, the old
Black Country vision of
hell-furnaces.

A lazy dessication. The soil first
parched, turned into sand, buckled
and sagged and split. In places
it would gape a bit, with soot
where the smoke came curling out.

And the gravestones
keeled, slid out of line,
lifted a corner, lost
a slab, surrendered
their design; caved
in. They hung their grasses
down into the smoke.

Strange graves in any case;
some of them edged
with brick, even with glazed white
urinal brick, bevelled
at the corners; glass
covers askew
on faded green and purple
plastic diadems of flowers.

Patchy collapses, unsafe ground.
No cataclysm. Rather
a loss of face, a great
untidiness and shame;

Silence. Absence. Fire.

Over the hill, in the lee,
differently troubled,
a small raw council estate
grown old. Red brick
flaky, unpointed,
the same green grass uncut
before each house.
Few people, some boarded windows,
flat cracked concrete roadways
curving round, and a purpose-built
shop like a battered command-post.
All speaking that circumstance
of prison or institution
where food and excrement are close
company. Concrete, glazed brick
for limits. A wooded hill
at its back.

Silence. Absence. Fire.

New Diversions

1
Dawn birds at noon
on a still
pale grey
April day with a tractor
behind the wood and
aeroplane sounds
hidden in the sky.

What light there is
sets them off,
even the single magpie
hung like a parrot, tail-heavy
in the sycamore.

The ringdoves are rising, rising
out of the gardens,
heads high and startled
between phoenix wings.

There's something in the fluid air,
something in nothing,
in the null
middle of the day.

They are in it and I
witness it. In the channel of
air between my window and the wood
the finches are jumping like
lice among the bushes.

The crest of the hen blackbird
ferrying debris to her nest and
perched on a post halfway
is a luminous emblem,
a great head
of last year's silvered Honesty
springing high up from her beak.

At a certain height
almost in the cloud
the rooks marshal and manoeuvre
in the something they sense.
They are in it, I
witness it. It is in
me.

2
Wanting to get away
from days when the taste of burnt sugar
seems never to leave; blunder
to London, lie frog-flat under a black pall;
wake in the after-smoke,
a huge bee at the window.

3
There's a belief to fix. It feels strong,
it feels far off. There is faith
to defile by fragmentation, so as to fit
into half faiths or less.

4
Mounted on all four walls, the neat
wide windows, framed in white,
idly take pictures of the weather
over my shoulder or
out of the corner of my eye, then
play them long after:

up there against the long desk,
watching the snow
come into Staffordshire
by way of the Cheshire Gap and Crewe;
how the flakes topple
softly among the low dark branches,
patterning faint passing shadows through
the silvery veil of net on the pane;
and how they will do that for most
of a dark afternoon,
settling hardly at all.

5
Vigil
taking over hours and losing them
into a moist gleam,
a single light

6
The eye ascending
is gone up
under the rafters
to a level where the black
screen stands as a trap
for likenesses, its tall panels
angled across the floor; images
flown into its darkness
and there stuck fast,
paralysed by the eye,
the eye that's paralysed with likenesses,
photographed, lithographed,
etched, invented, diverted;

in the middle of the floor, a sideshow
facing six ways, the images
flowering outwards from fixity
in the intense
black to spill through shadow.

7
Klonk! The smoker's trick,
foot in the door, acting
thick-skulled and frank,
asking for too much quickly, without
thinking of the risks of refusal.
Everywhere sick with old smoke.

8
Half hearing, half in a veil, to lope
and blunder, wild and dull. Medical lights
in a glistening inflamed ear. Cochlea,
balance, half-circles of canal.

9
I work in my garden
with bitterness

not strong, nor resentful
but baffled:

the flowering shrubs are in brilliance,
the vegetables
ready for sowing; and
a barberry I've reared ten years
almost to my own height

stands in the border
hung with bright apricot
blossoms over its dark
shiny leaves. I
wanted it and I have it.
This is the week it flourishes.

This is the week
I'm wanting no such colour;
I'd rather have grown, or found
a hazel copse
or a young birch wood
over Dane river,
and company.

10
The idea of defence
has been carried to the lakeside
and shown the sun. Over
a season of months it has unfolded
towers with hats, spiked snouts
and arches down to the lake shallows.
A system of optical levers works it
so that it can lunge
by a little way forward
or back, or edgewise, regroup
its bulk, and hunch; depending
which way you look at it; in what
weather, what mood.

11

Half reconciled, half healed: but
two different sets of halves that won't match.

12

If I declare *Peace*
in the middle of my mood

everything I can see slews round
slanting and sulking
against the new barrier.

13

A burnt year. Trick riders blunder about the concrete,
aimless, with no idea.
Open season for old wounds, odd jobs,
travelling from the one door to the other
over and over,
hamming it up, on the booze.

14

Out in the early morning, by empty roads
to a hump-backed bridge. In a book of
marked megaliths, the spirals
of the snail shells I found there in the dawn,
scattered in the stubble of the charred canal bank at Swanbach;
some shells burnt white, others still
patterned, a glisten of mucus left at their openings.

15

Rust. Limewash. The iron tyres
softened and pocky with orange,
whitened with powder, the oxide
bleeding out through it; all
the spokes, bolt-heads, coiled tubes
of the little horse-drawn sprayer
calcined and pale in the meadow, its

cylinder belly-tank slopping a stiffening whitish
puddle in the grass by its back end. Fetish.
Invented engine, trundled in, hoary
almost at once with its own
white slime; but with corrosion,
mark of temporary theft in this
age of iron, quickly
calling its forms back in. The air
drives its long waves
over beaches of discarded shipping, mile
upon mile of factories of rust.

16
In the evening the smoke blows far away;
there's a pressure in the dusk, beside
the swan's glistening canal at summer's end.

17
A counterfeit bucolic scene
tipped with irresistible acids
secretly slips into its place,
an unregarded patch of the exemplary
normality of Achilles's shield. In a well-lit
whitewashed factory, a herd of women
in overalls and mob-caps sit
placidly, merrily fusing
howitzer-shells of bright brass. The herdsmen
stand watch or move seriously among them.

18
The downpour that subdued everybody and
darkened the early evening dries out
into a huge exuberance;

in hard-blown sun the uplands above
the used-up quarries and burnt houses
are almost white, not
tufty with scrub, nor yet quite barren; the year
sways past the edge of a quarry.

19
Masterpieces in my sleep. A suppressed
novel by John Cowper Powys,

the core of its mystery
a high green mound that covered
legendary rumour.

At its climax
old Powys with his one
visionary eye
raised up a beam of energy
that blasted away St
Alfeah's, or whoever's, Tump
with a great cry and revealed for an instant, Yes!
the buried church, complete, the lost chapel
and the mound
within the mound; all seen
clear for first and last
as they vanished. For
the sake of the transfiguration
he annihilated the evidence
beyond all commentary. Maybe
that page of the text itself
never had empirical existence. I talked
in the dream with a divine,
worldly and humorous, the prime
scholar. He couldn't even *mention*
that apocalypse; it was something
deeper and more frightful than an embarrassment.
But we spoke easily
of the round grey nondescript mere
the author had left undisturbed
right through the action
in a dull meadow off to one side. 'It's
the only true *Pool*
in the whole of our literature!'
cried the scholar. Both claim and pool
seemed still to be
there to be agreed with.

20
In a dull hour thinking of the swan,
its white streaked with orange
rust; and suddenly to be surrounded
by the wind that blows through the door.

21
Duchamp and Man Ray
playing at the chess;
spotted in the café, in
a corner of the circular coffee-grinder,
flat wall
painted to a screen of angles

deep in the peace of the fetish,
of the portable engine.

Magritte Imitated Himself

I looked out of the window
once too often
(you'll *stick* like it)
and on a cold Sunday
afternoon of the Seventies
in Birmingham
stood, for the first time, disaffected,
on the aerial concrete
approach-platform of the new
Library, reared over the ghosts
of Widman and Dodd, Civil
Military and Ecclesiastical
Tailors, and Mason College, the old
Arts Building. Glancing to one side
I saw a skyline of certain
venerable cornices
in form of a frameless window
printed on the world.

Lost window, persistent world:
the place where I stood in the wind
was a sort of same; the space
where my customary seat in the first-floor
English Theatre used to be. They had
torn down all my support, removing
the very street beneath, then
raised it somehow up again, that
my attention could once more
wander. Starlings
used sometimes to fly about during lectures
and would look
ill-at-ease, time-travelling.

Provision

The irritations of comfort —
I visit as they rebuild the house
from within: whitening, straightening,
bracing the chimney-breast edges
and forcing warmth, dryness
and windows with views into
the cottage below canal-level.

For yes, there's a canal, bringing
cold reflections almost to the door,
and beyond it the main line to Manchester,
its grid of gantries pale
against the upland and the sky;
there's a towpath pub, where the red-
haired old landlady
brings up the beer from the cellar slowly
in a jug; there's a chapel
next door to the cottage, set up
with a false front and a real
boiler-house, and —
rest, my mind — near by there's
a small haulage contractor's yard.

Everything's turned up here, except
a certain complete cast-iron
housefront, preserved and pinned
to a blank wall in Ottawa.

This comfort
beckons. It won't do. It beckons.
Driving steadily through rain in
a watertight car with the wipers going.
It won't do. It beckons.

A Poem to be Watched

Coming into the world
unprepared

and being then always —
in honour of that
birth and to stay
close to it —
under-provided

and driven to exhibit
over and over again
unpreparedness

habitually
unready to be caught
born

The Running Changes

Driving northward in February once
on the run, to be clear of the Midlands
in a panic and ruin of life,

I heard the telephones ring in the air
for the first hundred miles.

But in the afternoon rain I found Sedbergh
and threaded on through it,
a silent close stone lock
which let me pass but barred my trouble;
I feared only it might be gone on ahead
to lie in wait for me by the Tyne. Then
the look of the road up to Kirkby,
the plainness and dark of it, settled
my stomach; and the sight of Brough
Keep, black as could be, risen in the fields
by a change of road, made me for that day
my own man, out over cold stripped Stainmore.

Another year,
coming down in peace out of Durham
in a late snowstorm towards sunset,

I met the lorries, headlamps full on,
thrashing their way up over Stainmore
in spray-wave of rose-tinged slush,
cloud-world behind and below them
filling the valley-bottom,
rolling, shot through with pink,
in the side-valleys breaking apart
to lance the pastures right across
with sunlight from no sure source:

and under the last trail of the cloud,
the vanishing up of its blush
into the grey, and the snow thinning,
there, once again, was Brough Castle
marking the turn southward,
and being dark.

Don't Ask

for Edwin Morgan

In the dream, the message
comes clear: *Hugh
Charnia has married Skellatis,
the Mari Moo.*
          Beyond that,
nothing. Just waking questions,
the dream draining
quietly away to leave
only an impression
of bony islands.

               Who are they?
Is Charnia the adventurer his
name suggests — look of a round-
faced surgeon or auctioneer,
blue eyes, grey flannel?
what's in it for him? Was he drunk?
Is he after his death, then, or his
coming-to-grief, that he
tackles the Marrying Maiden?

     *Undertakings bring misfortune.
     Nothing that will further —*

Skellatis with downcast
inward-turned eye
and long, sallow
crossed-over thighs —

             Is it
unknown energies by mystical means
he's after? What
business has failed him?
What title to what? *Could* she have
declined? does she possess
language to decline with? Does she at all
possess mind, as we know it?
Has he married above himself, then,
commoner of the same society where
she's the nobility? Or is he out of
a fatter one that overrides? Might she be

just an artist, whose pleasure it is
to be married mad? Or is she fly, and wiser
than he, and he an infatuate,
a colonist?

        If
she has the measure of Hugh Charnia,
have I, though? Was the wedding
really a wake
for one or the other? And
what *was* the toast?

The Lesson in Composition

Often it will start without me and come soon to where I once was
whereupon I am able for a while to speak freely
of what I have seen, imagined,
suspected, smelled, heard. I have never chosen
to speak about what I have
myself said, seldom of what I have done.
Though these things are my life
they have not the character of truth I require.

What I have been doing in the world as long as I can remember
is to witness and make conclusions. These are things
you cannot learn unless you dissemble, especially
if you start young. Like those of a spy
my words and actions have leaned to the oblique, my troubles
to the vague and hard-to-help. Likewise
I have been a teacher, I have been an accompanist.

Often it will start without me. More truthfully
other than without me it wouldn't, I have to be away:
for a while I must seem to be away, yet after so many years
I still can't pretend to pretend; taking that walk is compulsory, for
there's something about me
I don't want around at such moments — maybe
my habit of not composing. I could feel slighted,
knowing my own work hardly ever mentions me, except
by way of some stiff joke like this one.

— Tedium of talking again,
or at last, about composition and art, while I have one
eye on a thrash of clouds breaking around the guileless
blue of this December noon and the other
on the notion that there's no other topic to be had.
Whatever I start from
I go for the laws of its evolution,
de-socializing art, diffusing it
through the rest till there's no escaping it. Art talks

of its own processes, or talks about the rest
in terms of the processes of art; or stunts itself
to talk about the rest in the rest's own terms
of crisis and false report — entertainment,
that worldliness that sticks to me
so much I get sent outside
when the work wants to start.

I'm old enough to want to be prosaic;
I shall have my way.

The Passive Partner

'Just tell me straight: did you go to bed with him?
I mean, *did you have sex?*'

                         'Well — sex

was had.'

           'Oh God.'

'But — ah — very heavy weather
was made of it. Climax
wasn't reached. It's
been forgotten until it was
raised by you just now.'

Self-Portraits and their Mirrors

Harsh in the face
caught in a double
ice of daylight, a
mirror beside a window

Face
destroyed by the affront
in its own enquiry

The eyes staring at the eyes
staring and the features
actively scattering
away from that one
failure as if
suddenly caught
in hypertrophic growth
with no face to belong to.

Mystery Poems

1
Sloping and tapering away,
form of a dog's head, slick
to the eye but dry and
silky, polished with graphite-
black all across the prevailing
soapy khaki; and grown
from around its rocky shoulder
the side of a red brick
town after an afternoon shower,
the painted old iron frames
wet and a blind greenish-pale
reflection all along
the upper windows of Burton's.

2

Irish cheekbones and close-set eyes,
but not Irish, a man
sits at a loss in a farm kitchen,
looking at a higher chair as if
at a Labrador that's sitting on it
and knowing all the answers. But
only as if.

3

The upper corner — probably
battered concrete — from which
it all depends, is a comfort
by being in a form of order, and
boundless terror through being
inscrutable beyond the top of an upright
and the beginnings of a cross-beam
somewhere in the sky. Whatever
it is that takes this corner
as given
is infinitely flickering and free,
being shreds of primeval
dawn sky, patterned
by the moving parchment-coloured
twists and tatters of a square-rig sail;
blazons and insignia are
breaking away at speed and reforming differently,
as is peeled skin
picked at by curious hands made
in a painted-wood-sculpture style that
seems alien, even when they're gone.

The Elohim Creating Adam

Blake drew the guilt of God,
showed him at his compulsion
in plain view. Past the moment
of holding back and not
forcing the universe
to break into matter from the void;

by the very act obliged
for a while to be
imaginable, all the world
hurrying into being
and looking on

Blake, on behalf of Adam,
put the secret question everybody has:
'How was it, that particular
minute when I was made?'

And stained Eternity
with common answers: 'Troubled.
At cross-purposes. Recklessly aware
as to consequence. Streaked through
with the sense of something
suddenly and forever lost.'

News for the Ear

On a kitchen chair
in the grass at Stifford's Bridge,
the cataracts
still on his eyes,
the poet Bunting
dozed in the afternoon,

bored with the talk
of the state of literature that year,
sinking away under it
to his preferred parish
among old names, long reckonings;

but roused at the sound of good news
and surfacing with a rush,
a grunt of delight
from centuries down: 'What?
Has the novel blown over at last, then?'

The Nation

The national day
had dawned. Everywhere
the national tree was opening its blossoms
to the sun's first rays, and from all quarters
young and old in national costume
were making their way to the original National
Building, where the national standard already
fluttered against the sky. Some breakfasted
on the national dish as they walked, frequently
pausing to greet acquaintances with a heartfelt
exchange of the national gesture. Many
were leading the national animal; others carried it
in their arms. The national bird
flew overhead; and on every side
could be heard the keen strains
of the national anthem, played on
the national instrument.

Where enough were gathered together,
national feeling ran high, and concerted cries of
'Death to the national foe!' were raised.
The national weapon was brandished. Though
festivities were constrained by the size of
the national debt, the national sport was
vigorously played all day
and the national drink drunk.
And from midday till late in the evening
there arose continually from the rear
of the national prison the sounds of the national
method of execution, dealing out rapid
justice to those who had given way
— on this day of all days —
to the national vice.

The Toy

for Robert Graves at Ninety (1985)

Low arch, its marble overlip
by passage through
polished and moulded back
to a tallow sheen
touched by captive
daylights, their patterns to and fro
tugged in a membrane
that answers to leaf-shutters restless
on constrained branches at an aperture
or to captive water
set rocking in a cistern
within this wholly enclosed
fragmentary labyrinth
whose low arch glances
down to the inner floor below
and nothing more;
haunting is artifice, and a spy's view.

To be shut in
not with a disquieting
spirit that makes or
motions with the lights
but with the certain power of shaking
the whole of the place by taking thought
and the compulsion
not to resist the thinking of it,

the craning to perceive
sidelong, by moving sidelong
on a tilt, any way that it would,
and still to be shut in;
captive desire, set rocking
about an axis that locates itself
back, to one side and far above
in darkness.
            Shut
in, but perceiving in a glimpse
without grip, focus or tenure
the first of the concealed
spaces: nothing continuous with the marble
or its limited vista; but night air

all around and the rim of an old
galvanized bucket to touch; a jacket
thrown down but seeming
familiar; a sunk garden
thick with growth; moonlight
strong enough to imagine colours by.

On the Neglect of Figure Composition

PRELUDE

Hundreds upon hundreds of years of Lapiths
versus Centaurs; Sabine women abducted
by ferocious male models in nothing but
helmets and little cloaks, the corners
curled on their privates; such fading
mileage out of archaic feuds. It all passed.

But I propose a fresh Matter of Britain,
the parties to be as follows. First,
all those who have come to believe
our profoundest guidance to be
the person and style of His Late
Majesty King Zog of Albania: yes, Zog
of the white yachting uniform, of the pistol
for shooting assassins,
of the plump
Queen Geraldine, in matching
rig, and sharing his peculiar
gift of being able to make himself appear
rancid when photographed.
                                            All British Zoggists
mimic his dress and demeanour, his small
upturned moustache, the proud
suspicion of his eyes. In times of truce
they hold conventions at weekends
in the hotels of our former
manufacturing towns. They dress up,
eye one another, make wild plans.

Most repugnant to them are
the Ianists; once no more than a small
quasi-theological dream, but lately
more numerous, and moving in the land —
though not, it must be remembered,
given much to travelling. For an Ianist
is far more likely to project, or simply telephone,
his outlook and will, for implementation,
to another in the desired locality; they are
a sort of conceptual cavalry.
The heart of Ianism

lies in a constant meditation upon
The Real Ian. The Real Ian
is neither sportsman nor entertainer
but a part-time polytechnic lecturer
called Trevor Hennessy. He teaches
at two polytechnics, one
in the Midlands, the other in the Home
Counties; and he is not
explicit. Ianists are the most implicit
of all known people.

SKETCH FOR THE FIRST EXHIBITION OF THE NEW
HEROIC ART

'Ianists and Zoggists Resting between Engagements, in Rocky
Terrain'

'The Spirit of Queen Geraldine, Borne on a Cloud, Encourages
Flagging Zoggists during a Skirmish near Burnley'

Diptych: 'Members of an Ianist Cell Brushing their Crests/Appraising
One Another's Crests'

'A Zoggist Cohort of the First Rigour Surprised by Ianist Irregulars'

'Five Ianists Scorning to Interrogate a Captured Zoggist'

'The Zoggist College under Snow'

'Ianists Driving Randomly-Coloured Ford Escorts in Formation on
the A1 near Peterborough on a Fine April Morning'

'The Zoggist Acceptance of the Surrender of Weybridge'

'Suburban Panorama Incorporating Zoggists Enacting *Enforced Exile in Reduced Circumstances* and Ianists Enacting *Unacknowledged Supremacy*'

'A Modest Zoggist, Borne Unwillingly in Triumph on his Comrades' Shoulders'

'Zoggists in their Cups'

'Condemned Ianists, Already Blindfolded, Exchange Comments on the Turnout of the Zoggist Firing Party'

'Zoggists Hiding in an Ianist Laundry'

'Ianists Relaxing with their Women. They sit Silent in a Circle Drinking White Rum while their Women Dance Quietly to Records in the Other Half of the Room'

'*The First Time in his Whites.* The Mother of a Zoggist Cadet Proudly Puts the Finishing Touches to his Uniform'

'An Ianist Foraging Party Mingling with the Crowds in Sainsbury's'

'Oxford Zoggists in the High'

'The Passage of The Real Ian through Purley by Night'

Near Garmsley Camp

Under great heat we're searching
the slopes above Kyre brook
for antiquities: earthwork banks,
moats, mounds. There's a need
between us to discover something.

We're in a strange descent; curving
through a young plantation, aspens
or white poplars, spaced leaves
on straight pale poles, stage trees,
a wood.

                    Beyond that, the track
baffles, turns into nothing or anything;

but best, at the bottom of the wood
a field-gate chained shut
and an unmarked meadow, thickly
hedged round, and floating above itself,
floating a foot above its own grassy floor
as a silky, flushed
level of seed-heads, lifted
on invisible stalks and barely
ruffling; a surface cloudy and soft enough
to turn the daylight;
                            except where
close at hand you and I can
stare sidelong through it and down
into the measurable depth of clear air
and watch winged creatures swim
high and low through the stems.

Thus far down, and seeming
further; translucent patch set into
what seems the opaque ground. Above,
the bright opaque haze of the afternoon
has hilltop trees, towers, telegraph poles
rising into it as if into infinite
distance;
                but visible for miles
a man stands sunlit and hammering
high on Edvyn Loach church steeple,
trespassing in the air claimed for spirits
by the stone push upwards, and giving
the game away; an entire man standing
upright in the sky.

THE SHIP'S ORCHESTRA

THE SHIP'S ORCHESTRA

The Ivory Corner was only a wooden section of wall painted white, at the intersection of two passageways. To the left of it was the longer corridor; to the right at once there was the washroom door.

Ivory Corner for leaning against, the white pressing the forehead, the wood's vertical grain flickering beneath it up and down across the horizontals of the eyelids.

Washroom door swings, has weight, has rubber silencers. Limbs overhanging it from the Ivory Corner get foggy, the elbow gone, winging; a hand spread on the panel beside it stays brown and dry and shiny.

Always the chance of meeting that walking white suit with a big orange on it for a head; the white yellowed a little, as if through some sort of commerce with urine.

Then it was her black (purple, juice) net dress, rough to the touch, things grew so big in the dark. Or lacquered hair, dry and crisp as grey grass. Want it to come away in handfuls, and she be meek, and satisfied, as far as that. Plimsolls, the smell of feet in a boy's gymnasium. Learn to live with it.

Merrett calls his saxophone a tusk. What shape is the field of vision the eyes experience? Its edges cannot be perceived. A pear-shape, filled with the white plastic tusk, rimmed and ringed and keyed with snarly glitters, floating importantly. Where? Against a high, metallic and misty sunset, the sky like Canada in thaw, and Billy Budd's feet dangling out of heaven five miles up, through a long purplish cloud.

Potential fracture of Merrett's saxophone: by stamping, quick treading, sudden intemperate swing against an upright. In section rather like the break in a piece of dry coconut. No, it would not be likely to bleed. Just the steward brushing up bits of powdered saxophone from the saloon carpet, and Merrett, if surviving, looking out to sea.

Behind the rubber-stoppered door, the birth-basins.

Then it was her back, so broad and curved and deeply cleft, doughy and dry to the touch, like some porous cushioning that could not feel. The desiccated hair, yes, distinctly loose; all my senses precarious. I thought of the sheets as black, all hard things there are as ebonite, the indulgent back as very faintly luminous where I touched it; yet I was aware of something brusque in the air: a scented bonfire.

At times the sea rises uniformly to become much of the sky, harmless, translucent, golden-grey, with the great sun billowing down under the keel and flaking off itself from ear to ear. A wake of hundreds of scooped-out grapefruit halves.

Amy, too, in some of her moods, calls her trombone an axe. And the piano, whether I play it or not, is one of the kinds of box. Tusk, axe, box together joined. White baby grand box in scalloped alcove.

Janus, old door-god, your front face is alabaster, fringed with tooled curls, your cheeks and frontal prominences agleam; but a petrified, pitted arse, rained on for centuries, is all that confronts what's on the other side.

Dougal never actually speaks of his bass, even. But Joyce, the girl on drums, doesn't know too much yet. Judge the moment right and we can get her to call them anything. Tubs. Cans. Bins. Bubs.

A waterfall of orange-coloured deckchair canvas, from top to bottom as far as I can see either way without moving my eyes. And a long scroll — I can see the bottom of that, it is weighted with a short pole — covered with dimly printed instructions and transit data. Between these two, the projecting angle of two white-tiled walls at intersection. A narcissistic young passenger — I did not notice of which sex — has just left the picture, dressed for sunbathing.

Consideration of a porthole. Not punched or cut, but made by enormous controlled suction of plane surface away from chosen point of orifice; to be banded, clamped, bolted, glazed. For Merrett to regard the sea, his head resting as if provisionally on his small Napoleonic shoulders. Dougal has spoken to each of us in turn, to say 'Four days at sea, and they haven't asked us to play.' I believe he has also written these same words in a diary, the only entry so far. Dougal concerns himself a great deal with this question of our status, and Amy at least is beginning to be suspicious about his musicianship. This may be because, however obscurely, Amy is American, and is plainly a negress; being black, stringy and big-mouthed, although she wears her hair straight, while Dougal is equally plainly a late British Empire seaport (Liverpool) Spade; tall and medium brown, with quiet eyes and cropped ginger hair and a neat moustache of the same colour. There isn't a leader in fact; we're just a Foster Harris orchestra and if the ship people get any trouble they just wire the office behind your back. But Dougal has to bother.

The white suit with the citrus head ambles by, negotiates the steps with care. It seems benign today.

The taste of the first mouthful of whisky is a thing that creaks, like straining wood, but doesn't quite split.

About five of us, then, and something of an assortment. The coloration problem touches Merrett and me more lightly, in that we are, fairly decidedly, Caucasian, although I can tell already that there's a need for one of us to feel Jewish at times, and we pass this rôle back and forth tacitly. I am sallow and fleshy, with something of a nose, while he is more ruddy, with black hair and a pout. Both of us come from nondescript families; both of us are called Green. He is a Londoner. Both of us are circumcised, too; but so, as it happens, is Dougal. The other oddity is this Joyce, from Nottingham, who looks very young. She must be about seventeen, but doesn't look it: little face, rather pasty (has been sick, though); long blonde hair she can't quite manage; longish nose and big (relatively) dark eyes. Round-shouldered; sometimes a bit damp-looking under the arms. She hasn't unpacked her kit yet. Cans. Bins. Bubs. All five of us double violin.

Think of what all the people you see taste like and you'd go mad: all those leaping, billowing tastes through the world, like a cemetery turned suddenly into damp bedsheets with the wind under them. So the possible taste of a person is a small thing, just a flicker of salt, putrescence, potatoes, old cardboard across the mind, behind the words, behind the manners. And the actual taste, if you go after it, is something that's always retreating; even if it overwhelms, there's an enormous stretch of meaninglessness in it, like the smell of the anaesthetist's rubber mask in the first moments — it ought to mean, it ought to mean; but how can anything mean *that*? There must be a taste about me that could be sensed by others. Somebody as skilled as a dog could recognize it as mine; yet I cannot. If I try to get it from myself I just get the double feeling of tasting and being tasted all in one, like being in a room with an important wall missing. Hold hands with myself as with another person; the hands disappear from my jurisdiction. Looking down, I see moving effigies; the hands that feel are some way off, invisible. There is an image of me that I can never know, held in common by certain dogs.

White wall goes up to a white iron ceiling with big rivets. Windows higher up for a bent gaze. A grey canvas sky with the smoke streaking back from the funnel. This is like those afternoons on shore when everything seems to exist for the window panes. Somebody drumming on the grey canvas roof in my head.

Furniture all over the bandstand and the dance-floor still. The captain was soothing to Dougal, said, 'I can read music very fluently myself, you know.' If you gutted this little white piano here, sealed it and caulked it, it might float.

Joyce's hair by her ears and jutting over her forehead; her nose; the slightly separate gazes of her eyes: something clawlike in all these, latent and neglected. She and the others have been talking and I heard her say she was good at gym, at school. Plimsolls, steamed windows; rhythm brushes in the desk.

It splits my head! The great green-glass snout of the sea, the liquid thruster, like an enormous greengage sweet, with bubbles of air in it and the trails of sharks. Presses down into me, through my skull into the back of my nose and throat; peppermint, novocaine, cold and dumb. The sky, chalk-blue, squats over it, shaking, pushing. How did I come to be so far down, how did I come to be beneath the ship, to be like a figurehead embedded in the keel? In the flesh of the whole of my right side, from scalp to ankle, there is growing a wet chain, caked with rust. It's not painful; but when I move I can feel my flesh shifting minutely against it. Its tension is different from the tension of my flesh. The old schoolmistress sits at her high desk in the chair with its own footrest. Beside her is a big extension loudspeaker with sunset rays across it in fretwork. Piano music comes from it, and the beating of a tambourine. The old schoolmistress sings, swaying in her chair. The lights are on in the classroom. The little boy lies on the floorboards.

A person is a white damp thing — and here's Amy, who's black and dry — a white damp thing, greyish in some lights even when alive. You could inject salt water into the human body. An all-over emetic. Ha ha. Seminarist. Plankton. Bathyscape. Handkerchief.

Impossible to believe the sound in a piano is so far from the pianist's fingers. I know the keys are ivory boxes filled with wood. In the key of E they seem filled with the pulp of teeth; in G with butterscotch. When I play alone the music is never without a voice or a body.

Lizards, we are all lizards, or will be: khaki-green rubbery lizards prancing agape on a plinth covered with plastic sheeting, over which the cold water is kept running, out of respect for our nature. And we shall not feel sorry for one another when the blunt scissors jag at us and the cold fluids trickle sluggishly out.

Old man up on the boat deck in the morning sun. Sheltered from the wind, wrapped into his chair with a rug. Flattish hat, with the crumpled brim turned up; muffler, sunglasses. Little old man with a clean brown monkey face, mouth like a sloppy purse, livercoloured lips; hands spread out, spatulate, fingers pressing into the rug. He said to me: 'You are a pianist. I am a masochist.' Merrett came up and I said, 'Old man, we are going to pick you up by the kneecaps and throw you overboard while nobody is looking.' 'Gentlemen,' he said, 'I am not a homosexual; you misunderstand me.' Etc.

Is it good to feel, under the skin of the chicken as you hold it from running for a moment, the muscles you are going to eat? Oh, questions, questions. How can you crucify a man with a giant orange for a head? The orange falls off once the body slumps down between the strained arms. The shirt collar feebly tries to mouth the last words; you replace the orange, it falls again. You can't put a nail through the orange to hold it to the cross; that's another story. Stand holding it up all afternoon, and the shamefulness of the detachment in the dusk, when he's cold.

A huge yellow oil-drum afloat in the waters of the bay. Sunlight.

Throbber, she said, you're my throbber. And you're my gummy, was the reply. My guggy; my guggy gummy. Now you're my thrubber, she said.

I have known this all along.

Astringency, the prickling of the scalp, flexing of the feet, rotation of the wrists, passing the hands round the confining surfaces of the room where one is. That done, the thought of the scalloped alcove where the band might play. Combed plaster in swirls of rough relief, a deep pink rising from the floor to meet the powdering of gold that thickens and conquers at the zenith. Floor projection forwards, a curved apron, no higher than a tight skirt can step up from maple to black linoleum pitted with marks of casters, drum-spurs, bass-spikes.

Swung from the arms of the gaslamp that was the only light in the street; a street greenish black, among factories. The long linen sack was twisted round and round and was unknotting itself in slow revolutions, with all the weight at the bottom. As it turned, the moisture caught the light, coming through the fabric from top to bottom, but not dripping. Kick.

I am in a poster. This is how the whole thing's meant to appear, obviously. Somebody has been at work on my perceptions, cutting them as giant rudimentary forms out of very thick softboard with a fretsaw, and painting each one a single colour. Although they're only two-dimensional shapes they're thick enough to stand squarely. The ship's superstructure, away over me here in the sunlight, is huge and straight, and of an immensely comfortable white. It's not at all complicated. People, the sky, the sea. The cataract of orange deckchair canvas, the scroll; these are present, and a march of relief-built letters you would need a ladder to climb. An E, an A, a T. You don't have to go and eat. This *is* Eat. Scaffolded, boarded, painted. This is the provision, this is the activity itself. Maybe I have hands a yard wide, a smile like an excavator, nothing matters. The dimensions of the components are not determined by the component subject. Fine. The directors of the shipping line, Foster Harris, the captain of a distillery can be seen queueing by the ballroom door to take credit for this ordering of things. They are small and neatly photographed against the placid outsize expressionism of this set. Then there's the sun. This is Eat. It says it is, so it is. Things seem what they are, believe it.

Looking at this world that is like cake, this fifth day at sea, I realize that Amy, Merrett, Joyce and Dougal are probably happy people, to whom a day like this is nothing strange. And I was thinking of them lying down below on their straw, sniffing and shifting about. Joyce on her straw. Merrett on his straw. Joyce with her clothes too thin to ward off prickling, Merrett with nowhere to put his glasses and his trousers too tight to lie down in comfortably. Dougal on his straw, lying on his thin shoulders and knobbly buttocks, scratching, scratching, his long fingers always squeezing at his skin. Amy on her straw, hard and glossy, waiting for her belt, her dress, her skin itself to split under the strain of not caring. In fact I know Dougal and Joyce are playing draughts, Amy reading magazines, and Merrett lying benign on his bunk playing cat and mouse with a hangover. Snug little figures in the big poster. We are getting to be like the passengers. They should let us play, perhaps; treat us in some little way as if we were a band of musicians. Little way, big way: the dimensions of the components are not determined by the component subject. It doesn't matter. I see, coming up a stair, through a door, round a corner, up an open companionway across a deck, through a door and out of sight, going, the actor who must be going to take the part of me in the immortalisation of these days — their 'rendering'. Bigger than I would have expected, and a bit old-world. Tuxedo, black hair, suntan, high-powered eyebrows, searching brown eyes; boyish manner preserved in maturely male bulk (shruggable shoulders, big back). Glasses in breast pocket, presumably. Do Merrett, Dougal, Joyce, Amy, see their actors and actresses today? Probably not. Why did I have to see mine? I didn't want to spoil the poster by appearing as myself. Why not? This is Eat. Take what comes.

Throwing up in the washroom the other day I had a vision of a dark pink, double-tailed mermaid. I haven't been seasick this trip, so far as I know: this was the little drunk I went on with Merrett and Amy to get settled in.

There's a binary phase in this kind of vomiting, especially marked if your balance is fairly good. A strong consciousness of two ears, two shoulders, two knees, feet, elbows, sets of fingers gripping the edges of the basin; these two sets of characteristics existing each on its side of the room. Between them is a void, a gully; and that is the vomiting. It was in a sort of clear sky above this gully that I saw the mermaid; just at the moment when the idea of being sick rises to the ears, brims and fills them like a sea, the sight goes and the sudden assault on the pharynx arrives, and the invasion of the facial expression. She was floating Botticelli-fashion against a greenish watery sky and some way up from some very stylized olive-green waves. The two tails, in obvious concession to the binary thing, pointed to left and right, and were, at the extremities at least, green as the waves. For the rest, apart from some rather nondescript-coloured hair, she was coloured this remarkable deep pink, uniformly, without variation of tone: lips, nipples, fingernails, the one eye she had open, everything. It was the pink of scouring paste or a rather sickly tulip, a little bluish, yet very bold. Although clearly breathing, and even moving a little, she looked like a figure in a primitive painting whose artist, while realizing that flesh wasn't white, hadn't got down to details. She was a burly girl, with fat rubbery cheeks and round arms; looking out of her left eye at me as though she had never seen anything like me before.

Amy has begun to play: the first of us. I can hear it through the wall. She has got out her trombone case, removed the instrument, and is blowing it. Long notes, staccato series. Methodical, clear, accurate; says nothing. Amy is a killer, a musical shark. For those who want that kind of treatment. She'll not bother with me. She must be feeling low, to have to play.

Soon there will be a meal. The food will pity me, I shall pity myself. Healthy, ambulant, I am about to be fed with cosy food that tries to make up for my being far from home, my being a great big boy criminal. The seat will be soft, the things clean, my last mess mopped and laundered. And until that big soothing spreads towards me the little notes of the trombone hammer away, like brass shell-cases on a moving belt. It is the sort of time when something very large and wide and silvery, like the capsized hull of a vast ship, can begin slowly to rise above the horizon.

Reasons. The ship is a unity. Enclosed within its skin of white paint it floats upon, and chugs across, the unified ocean. Some would think of it as having the shape of cleavage, a narrow leaf: to me it is a flat canister bearing another canister and a similarly cylindrical funnel, the basic canister shape being eccentrically elongated. This is because the vessel's speed is not great and, whereas there are those who would see the superstructure as a vague and mutable spectre above the hull, it is that hull that appears ghostly to me, while the funnel never altogether leaves my thoughts. At any rate the ship is a unity and does one thing: it proceeds on its cruise. Not only does it have a structural and purposive unity; it has a music which proceeds with it, sounds within it and makes signals of the good life. In among the musicians is the tough glass bubble of the music. Reasoning, now. The musicians don't play. No bubble. The ship is not a unity. It is not white. It is grey, indigo, brown. Thin girderworks of green, and orange even, and coils of pale yellow piping. It is not a series of canisters; it is a random assembly of buildings which, though important-looking, have no proper streets between them. It does not float; its parts are arrested in their various risings and fallings to and from infinite heights and depths by my need for them to be so. The funnel cannot be said to crown the firm structure; rather it juts rakishly over inconsequential forms and looks when the sky is dirty like the chimney of a crematorium suspended above the waves. The ship does not proceed on its cruise, but opens and closes itself while remaining in one spot. The ocean is not a unity but a great series of shops turned over on to their backs so that their windows point at the sky.

O captain. Is it the captain? O first officer? Is it the first officer? Etc.

Such heavy straps and buckles for so young a girl to wear! Such a stiff casing and mask, such mechanical magnification of the voice to stridency! Such a channelled street, with iron pavements for her to strut down, so young!

Monitors, those curious warships there used to be. Little vessels that each carried one enormous gun. Restless home lives of their captains.

The rings of winter, the circles of winter. Why? The hoops and bands of frost. Cooperage that fetches the skin off. Why circles when it goes cold? There are times when you can live as if in a round pond, keeping on moving even when it freezes. And overlapping ponds all round, across the gardens and the streets; making up the sea when the land stops. The rings are there but nobody can ever see them.

Think of Joyce's mother. An accordioniste, maybe, toothy, gilded somewhere, and with a hollow at her throat you could rest your nose in after a hard day's work. To turn her child into this, what can she be? Yet the girl thinks of herself as a jazz musician; talks about Blakey and Roach, or mentions them when pressed. Think of Amy's mother. Difficult. Think of Merrett's mother; of Dougal's mother. Of mine.

He was in a garden all walled about and set amid the sea. And he came into a place where there was a soft-faced flower like a cup on a single stem; the bloom a little larger than his own head and its top a handsbreadth taller than he. And soon the flower lay down on a low bed that was in the place and gave him to understand he should lie on the bed beside it. And he did so. Whereat the flower lay close with him and softly folded him in its leaves, as well as it was able. And he was aware of a marvellous scent from the flower, and would have swooned, etc. And forthwith the flower made great to do to unloose the fastenings of his garments, even to the buttons of his braces. And right hard the work proved, whereas the flower had not fingers but the points of its leaves only. So in this wise passed a longer while than that of all that went before.

The rings of summer for that matter. Carry on.

This is what it is like on the land. She: she holds court facing upstream, on a handrailed plank bridge over the yellow floodwater of a ditch cut into the clay to hold a gas main. The gas and the clay stink like dung in the cold; pink smoke jets from a vent high in an isolated building some way off. The place is surrounded by white canvas screens, damp and grubby. Over her waved dark red hair is spread a muslin dishcover with a bead fringe. Angry brown eyes, pasty skin in folds down to the dewlaps and scrawny neck. Head raised, wide mouth pulled into disapproval. She sits straight-backed in the old cane armchair, propping herself on one elbow. One leg is crossed over the other and the fluffy slipper points elegantly. Apart from the slippers, the dishcover and a pair of baggy pink drawers, she wears nothing.

Merrett, Dougal, it is you and I who have put her there; struggling in our leather breeches through the mud of the site, carrying her at shoulder height in the cane chair; Joyce, Amy, in dungarees and waders you were there too. We must be together in something. Far off across the wet land there are conical fires perhaps and men turned into meat.

The rings of summer would be visible if they existed. The powder we used to make orangeade from, cast in big circles on the ocean, and the circles widening and fading as the powder sinks, in curtains through the depths. And in the empty middles of the circles white things could rise, and float, and disappear. Whalebone spars, cakes of soap, plastic saxophones, tennis shoes.

I saw it from above at dusk as I looked over the rail. On the deck below, it sat hunched, the white suit full of blurred shadows. The head is larger, puffier, more yellowed and sad, and it shows indentations which have not filled out again and which seem to be the product not of blows but of violent fondling. I think the end will come, probably by further violence, during the night, or tomorrow at the latest.

My head. The huge shimmering cloud-filled canister that supports it by describing its furthest limits is shaking irregularly in the night breeze. There is light on the waves, and the ship is a dark factory.

Ivory Corner, white and shining phosphorescent, a tongue that licks me slowly as I approach, from toes to scalp, and extinguishes itself at my back.

I have talked to many people today. I have talked a great deal. The question of our not being asked to play has gone cold, even among ourselves. We are accepted everywhere as what we have become. People go off to bed early in the silence. The absence of music is somebody's urbane whim, and they respect it. Maybe there is somebody slow who will notice, tomorrow or the next day, and be indignant, just as Dougal was at first. But think now of those dozens of silent black aridities moving about the ship, going into cabins, losing momentum, sleeping, turned inside out in the dark like rubber gloves.

There's a labyrinthine system, running all through the ship, of whatever it is that rules by default the minds of the incurious. Slippery wet blackness, invisible by day. Sacs, coiled tubes of it, linking all the people, deck to deck, dream to dream.

I once actually met one of those men who say to you, 'As a matter of fact, sexual thoughts and activities play a very small part in my life.'

Pink smoke jets from a vent high in the wall. Ibis-coloured. In the yellow ditch float bottles of clear glass.

At sunset Merrett grew frightened and stopped drinking. He tried to tell Dougal and me stories about himself. We listened quietly, for I had not been drinking and the stuff seems to have no effect on Dougal. Merrett, in the red sunbeams, talked and laughed with us, while the coming of night alarmed him.

Then it was her black net dress, rough to the touch, and the warm dry scent everywhere catching the breath; and it was the grey desiccated hair floating and filling the room, the dress and the hair up to the ceiling, the room a skirt, the hair in the ceiling corners with the smoke; hair from her scalp, legs, belly. Useless to open my eyes, I was blinded with touch. But her skin was nowhere; the body was away. She had filled the room with her dress and the dead hair. Pouflam! The fire caught it.

208

She is among the hollyhocks, she is on hands and knees among the rhubarb, she is legging it over the low fence in the dusk. She still has the black dress. She still has the grey hair. I see it behind the petrol pump where she is standing, her back to it, pretending to hide. She is too mad. She doesn't really feel what happens to her. I insist that she must. Who made her mad? If she were really to go bald her breasts would become beautiful, etc. I have to be sentimental about her, for her own safety.

I wake, and ebony poles are across me from wall to wall, a few inches above my face. No farther apart than the bars of a baby's cot. There's a grey dawn light travelling the cabin; it goes. No. Sleep again, in this paper leaf. I have wet myself, I have died. No. In my sleep they have anaesthetised me and with their toothless rubber jaws they have gnawed away my genitals entirely. Cleaned me up — I hear one of them still mumbling it. I dare not touch yet. The grey light, the white light, the dull disc of waking. Not yet. In the night Amy comes to me now. She strips her hard black body, a piece of furniture, and presses it down on me where I lie as if she would break my bones. And with the expanding mandibles that have replaced my privates I clasp her and contain her sadism for hours without motion, until she lifts herself off, quietened, though still taut. Her straightened hair sticks up in a crest as she digs her fingers through it. She fastens her white towelling robe and gives me a dog's snarl of a smile as she goes.

Amy does not come to me in the night.

My actor goes past, treading lightly, his big shoulders affable. He greets me; I respond. He has useless-looking hands of course. Who will play for the soundtrack, will there be any soundtrack, etc. As well as his own breakfast, he has eaten mine.

Visceral pipes of white porcelain, huge things, in banks and coils, too wide to straddle. How to get lost in the morning. They reach up in stubby loops and descend far beneath, the systems crossing, curving round, running in parallel. Some plunge vertically through several levels then divide and disappear; some come creeping through the interstices of others as if squeezed before cooling. At the top, the light strikes hard and bright, softening and blurring at the next levels.

Deep down there is only the little light that drips deviously through the chinks. Every so often there is an open end, pointing upwards.

Potential fracture of the pipes.

Hope-pipes, love-pipes, fright-pipes, thought-pipes, loss-pipes, hate-pipes. Pipes of coarseness, pipes of sanity. Pipes of confession. Pipes of purity, pipes of sanctity, pipes of flight. Riding-pipes, rubbing-pipes, sliding pipes, wiping-pipes, confronting-pipes, adoring-pipes.

Potential fracture of the pipes. Virtually impossible. Only single-handed with a light sledgehammer, squatting on the topmost U-bend and clouting at the pipe until it cracks and shows the ochre stuff it's made of. Then bashing at the necks down to the levels, smashing across the fat conduits that curve down, caving in long sections of horizontal pipe from above, then standing in the channels and striking out at the sides till the brittle catwalk underfoot collapses; the débris all the while shuffling its way down in shards and dust into the open mouths and between the pipes to the bottom of the whole system. Descending, the need for clearing the rubbish from the pipes to get a foothold; the monotony of the straight stretches; the strength of the main joints. Arrival at the level where the débris no longer shifts, but has accumulated to this height up from the floor; and how far down that is, under rubbish and unbroken pipes, it is impossible to discover except by reaching down to it. Then the task is to probe for the buried pipes, shattering them among the surrounding fragments, never being able to clear their surfaces any more, but hammering at them blind through what covers them. The excavation of hollows to work in; the seepage back of the broken pieces down the slopes. The laceration of the boots and gloves, the sensation of the feet sinking deeply into the jagged shale; the pipeless walls of the system's container staring inwards at one another, feeling the new light down themselves towards what is at the bottom.

Then it was that man learned to fly. Unfolded from the middle of his waistband by the pulling outwards of a black cord was the creased brown paper bird-form, crackling in the sunshine and peeling itself out bigger across his belly and chest, pushing his shoulders back and flap-wrapping around them. The tail that flattened itself across his thighs; the paper membrane that stopped his nose and mouth, closed his eyes and clapped wind at his ears in immense distance. The railway that ran southward across the smooth ugly sea.

Dropping from the sky and going fast, a cone of paper or some composition fibre, white tipped with red. Disappears below my lower lids, behind whatever is there for it to disappear behind. Effusion of aeronaut; part of a trombone mute; spiritual part of man-made cat.

The paper membrane that stopped his nose and mouth, closed his eyes and clapped wind at his ears in immense distance. The hand that at the same moment walked its long fingers up the back of his neck and through his hair, prodding quite hard, treating his head as something unlikely to burst, letting the pointed nails pivot in the skin as the fingertips turned over. Joyce.

Joyce, who looks at people sometimes as if she lived in their bodies and had just moved a few feet away to get a better look. When her life invades her daze. Too soon yet for anybody to tell her.

Her looks, not her life, invading the daze. The life's not powerful enough to alter without the looks. But in that pasty little face the mouth is going to be wide, with an upper lip that pulls back at the corner. The brows will be thick over the irregular eyes and the nose long and straight, with a knobbly end. Some of her postures foreshadow these changes.

The mouth. Fills the area of vision, is very close. Soft, the woman's mouth, impossible to tell whose. Colours fade towards vapour. Closer, there is nothing but the lips, their joining line rising and falling along their shapes, the wrinkling in of the surfaces towards the line. Slowly the line retreats, the lips part, widen; the teeth can be seen, the gentle tongue, not quite motionless, the spaces of the mouth, capable of holding a clear note of music. The breath must taste of cold water. From the right Merrett walks on, the lower lip level with his knee. Seeing the open mouth, he peers forward; then, putting his hands into his jacket pockets, steps carefully over the lower teeth and into the mouth, ducking his head. He walks cautiously about, looking up, down and around; but does not go anywhere out of sight. He ducks again under the top teeth, steps carefully over the lower, and comes out. Then he goes on his way. The mouth stays open.

Somewhere there's going to be some music. I haven't the courage myself to clamber over what keeps me from the piano, to plunge my fingers into its clashes of sound. And what I play isn't what I mean by music. Breath music. Slow opaque music. The ship has come close, drawn itself up my body and continues to rise. Yet it is, though fitted to me, nevertheless very big and stretches far away from me above and below and on all sides. And all the compartments of which it is made are full of milky sounds ready to knock against the bulkheads and echo all through the vessel.

The captain's hat revolves, returns, revolves, returns, never completing a revolution. The captain breakfasts above the clouds, on thoughts invented for him by Dougal.

Great glycerine drops of water trickled down the girl's bedraggled hair, were caught in her eyebrows, ran down her nose and off its tip. Down her cheeks and neck, cascades. All over her chilled lumpish flesh, the big grey eyes looking down at it, the wide mouth curving, the tongue licking the drip in from its corner. A rivulet between the breasts, spreading across the steep belly. Twisted streams down each breast and falling from the cold plug-nipples. Water standing in separate shiny drops on her big thighs.

Corridor of grey mucus. A kidney bowl of it behind each door.

Over the white linoleum floor the nurse advances with the orange gladioli arranged in a spray, then goes off behind whatever she goes behind. A tap forcefully turned on. The nurse is mature, ladylike, and of fair complexion.

Dougal tranquil, nodding down there in a canvas chair, his cuffs rolled back, the gilt strap of his wristwatch gleaming. His cigarette, cocked between his fingers, burns down. Then which of us is worried?

Shirt cuffs folded back, right above the elbow, whites of eyes showing. Let's go to church to see the dog given its fix.

Merrett said last night that his alto would dent rather than fracture; and that he also had in his trunk an ordinary plated metal tenor, not so handsome, but capable, he said, of cracking a man's skull if you hit him hard enough with it.

Blue fog. Electric cigars. E-flat horns wired in series. *Camions.* One eye at a time, one eyebrow at a time. Dougal dozes. For Dougal to have his proper beauty the circumstances of his life — waking, going to sleep, washing, eating, defecation, micturition — need to be regarded as clinical conditions, their operations supervised by trained nurses.

The two little scrubbers hugging each other in Merrett's bed in the hotel, both of them snivelling and complaining; and Merrett, in shirt and trousers, lying rigidly under the covers on the edge of the bed. Scene from Merrett's first professional job, as clarinettist in a traditional band touring out of London. The bed, Merrett's and the banjo-player's; the idea, the banjo-player's. Nocturnal disappearance of the banjo-player.

All this disposal business, these basins, enamel buckets, plunging tubes, embalming sluices, constant jets, sterile bins, sealed incinerators, consideration of where the banjo-player might have gone that night, of the abolition of words taped to our memories, of the storage of one night under another night, the earlier ones gradually fading as the multi-track builds up beyond the bounds of desire; all this question of the attenuation of substance to concepts. Are there in the ship's mortuary yellow-soled feet with the toes sticking up and facial lines of resignation showing on them in their stillness? Where are our instruments? They luggage us, they follow us, they squat behind us when we're not looking. The facts are these. Merrett's plastic alto in case beneath port-hole. His tenor locked away in his trunk. Dougal's bass, in cover, standing behind the door of the small room behind the bandstand. Joyce's drums, boxed and cased, in the same place. In Dougal's cabin, two violins in fabric cases edged with leatherette and an acoustic guitar in a polythene bag. In Amy's, the polished trombone, with chasing all round the bell, its mouthpiece set against Amy's lips, while she blows long notes, very quietly. In the corner, a small guitar and amplifier. In my cabin a stack of folders of standard invertible reversible orchestrations, song copies and manuscript paper. The white baby grand piano, locked, its lid down with crates of tonic bottles stacked on it, in the scalloped pink and gold alcove. Disposal.

Potential utter disposal of the instruments, itemized disposal (as stamping on the violins, hammering the drums flat-sided, sawing the piano into slices like a ham, turning the trombone into an artificial flower, fighting a duel with the guitars, shredding the bass with pliers and chisel) being too crude and guilty. An engine is necessary: a hangar stretching some miles in all directions, with every part of it, outside and in, painted matt white. Semi-opaque panels to admit much diffused light. The instruments fitted into white foam rubber containers sealed laterally and set in further containers sealed longitudinally, these last being cylindrical and of uniform size, about ten feet in both length and diameter. As many as possible of these cylinders; all must be uniformly weighted, and each must have an identifying number stencilled on it consisting of a number of digits equal to the total number of cylinders, only one digit on each cylinder varying from the norm established in the first numeral, and the varying digit not to be in the same position in the sequence on any two cylinders. Filling the hangar, a continuous white tube into which the cylinders fit and in which they are moved pneumatically at a steady speed. Further details of solvent tanks, sludge filtering and caking, moulding of cake into casing of fissile explosive device, recording of distribution of post-explosive material, public opinion poll, suspension of communications for necessary periods, change of languages, etc.

Without the instruments: we can all share a taxi and spend the afternoon at somebody's sister's wedding party. Wellington boots, exuberance, ducks and drakes on the park lake with crumby plates.

How the sloping shed of a Saturday evening in England falls over Joyce, over me, over Dougal, Merrett, over sad Amy. We do not play, we are people. Dougal embraces Amy in a delivery van, Merrett and I go to a gymnasium just before it closes. Joyce is one of the little girls who giggle at us as we go in.

This is what it is like on the land: on ground where disused vehicles are dumped, a woman has given birth to a child in a giant aeroplane tyre.

On the land the oil-refineries strain to escape from themselves along the river banks but cannot move, and the sky on its conveyor comes round and round again.

On the land the men swarm over the new concrete obstacles and fill the spade's ravines with their ebullient bodies. Let us build again!

The ship's orchestra is at sea. Crammed into a high and narrow compartment in a heated train on a penal railway, we loom out of the shadows at one another in our full dignity at last, between the brownish light of the windows on either side, light that fails to reach right into the domed ceiling of the compartment. The light paints over Merrett's glasses and covers his eyes. Amy's cheekbones are luminous in the tobacco shadows; our heads reach up close to one another, preternaturally large from narrow shoulders and stretched bodies. We are about to agree.

Perhaps the little white piano has useless dampers, and however good the others are my playing will be a continuity of shining brass water, shaking idiotically. Have the others wondered whether I can play? Pianists who go about alone usually can. For my part I have seen Dougal stowing his bass behind the door; have heard him scat odd bars; I have heard Merrett blow a few sodden flourishes on his alto when he took it out to show it to me as soon as we were drunk; I have not seen Joyce anywhere near her drums, but I have heard her humming to herself. I have heard Amy's short notes, and her long notes; and what appeared to be a series of arpeggios of the chord of the fifteenth, with the fifth, seventh, ninth and thirteenth degrees flattened in various combinations as the afternoon proceeded. Some of them slowed her up a little, but it would have been an achievement even for a woman who was sober. Amy has stayed drunk in order to break Joyce in, it appears.

All the same, we're about to agree. The guards are laying the jumping chains out in the sunlit courtyard. The lizards scuttle for shelter. The weeds that have had links dragged roughly over them straighten up slowly.

214

The old man on the boat deck, sitting wrapped in his rug, turned his sunglasses towards us and, seeing us all together for once, suggested to Merrett, Dougal and me that we ought to dress as women. It took him some while to make clear what he meant, and when he had finished he sat laughing and laughing, his mouth open all the time. Amy was much amused too, and wouldn't let the idea go. It turns out that she is the only one of us who has actually been married.

The white suit going round the corner into the washroom, shuffling with a stick. The length of the corridor away, and the light bad; but the head appeared to be bandaged.

My body explored slowly by squares of differently coloured light. Odd sensation. The little slanted rectangles alter the sizes of the parts of my body they touch from moment to moment and leave a black creek of me in among themselves, that waves and shakes itself about in pursuit of them. The flapping black unseen part of me: a unity.

A covering of sacking over me, that I needn't move. Beyond it there is a room distempered pale green. Smell of soap. Inside the sack, in here with me I think.

The big standing dog. The silent dog. The chimney is stuffed up, the cracks round the doors and window sealed. The brown dog, motionless, grows to completion within the room. Nothing to consider there but its rough matted coat, its deep flanks.

The porthole of enormous strength has come among us and stands, turning this way and that to be admired. Adversative at all times but turning, turning always to make peace. The big standing porthole. The brown compartment. Our soap-sack, our own, the scent of which guides us back to ourselves in the night. The dog's huge stale smile, its mountainside of coat; the silence of its smile.

Hear the lining of the chaps part here and there from the gums, the saliva and air making sound in there. Sound that peels back behind the bulk-heads, across ceilings; up and down the backs of the legs. Sound fitted with a glass eye.

Inside the wet aeroplane tyre: when there are enough strips of bacon rind we shall weave him a little coat.

Sound that moves beneath its clear brown glass, propelled by full sails; that carries reservoirs of ink beneath its waist; that shadows itself as it goes. Sound that swallows pearls into twilight the colour of beer.

215

Slippery sound, retarded till it fractures into many transparent wedges, then into countless pools of travelling light, within each of which great black streets stretch themselves before the windscreens of lampless vehicles, travelling fast. Sound that polishes the stinking dust and makes it stand up in the night, before the wardrobe mirrors, while the gloves deflate, crinkling in the dark.

The sound of hundreds of feet of film, split from their reels on to the store cupboard floor, being trampled in the dark by the animal shut in there.

If Merrett, Dougal and I dress as a women, become women, will Amy and Joyce have to become men?

There'll be no need.

In the domed compartment, so ill-lit, where the spilled celluloid crinkles out of sight under our shoes, Amy's knees touch mine as the train sways; Dougal's knees touch Amy's and mine; Merrett's knees touch Amy's and mine and Dougal's; Amy's, Merrett's, Dougal's and my knees touch Joyce's as the train continues to sway. If only we could all play together on one single instrument!

Ivory Corner for Joyce; on the white paintwork a big lipstick mouth to kiss her. Ivory Corner for Amy: padded hooks, to hold her up by the shoulder-straps.

Ivory Corner for Merrett: with a heavy iron disc to press down on to the crown of his head when he stiffens upward. Ivory Corner for Dougal: Joyce, standing stark naked and freezing cold, with her eyes shut, at two in the morning.

To be somebody else: to be Amy. Like this. The men push me out of the washroom if I go in there. Grabbing my elbows, knee in my butt, hand shoving my head forwards at the door. One pulls the door open and the others shove me against it so that the white edge of it comes between my knees, splits the skirt, strikes up my belly, and my teeth and nose come hard against it. This is to make the black ape-woman swear; the bitch whimper for her fix.

What do they think I want? The sea coming up my street, fast, kicking up my nose, making my calf-eyes roll at the sky, splitting me with an explosion of green glass foam? The rifles laid across me in a heap, where I lie naked, the cold bolts thumbing my rifle-coloured skin? To go around smelling of dead flowers again? I don't know what size of things I want.

Anybody, mummify me. It doesn't matter which of them I am.

The little shrivelled black monkey, growing smaller and smaller. The blue above: morning in Sky Gulf. The monkey's owner knocks nails into a piece of board to make music for it.

And the alabaster kingfisher plummeting upwards through the grey photo-print of the water for minutes on end, out of sight. Dedicated.

She's wet all over, with a thin film of something slightly viscous, almost like very watery cement. It must be dropping from somewhere overhead. Cold sweat out of the metal. Not so much on the hair though. Glistens: phosphorescent, the sheen wavering as she breathes. Her lips are lightly stuck together with the stuff, and the eyelids too.

Why has Joyce been crying?

The single instrument would have to be an inflated ring, like a tyre-tube. When it's wet whatever rubs across it raises a squeal. Each of us to have an undefined territory on the circuit. Not really a tangible thing even though it's our common body; but it's as if there was an invisible sphincter in the sky somewhere, with a fivefold answer to our touches.

Buckle the lamps in close to the ribs, rub salt on to the pale peaks of the shoulders, clasp stones wrapped in rag in each armpit; paint the mouth of the navel with lipstick. Pin newspaper cuffs round the nipples and the groin; whether it is a man or a woman is immaterial, but shave the legs and forearms in any case. Whoever it is will need to stand up throughout. Thin white steampipes run down the pink wall.

I am something that has been pushed out of Amy's body, though I cannot remember it. I have no legs, though I have the idea of legs, and I have no arms or hands, though I can conceive of them; but I can move my head this way and that, where I lie. She knows I have come out but she doesn't know where I am.

It would upset her very much to learn that I can move my head in this way, and I shall take care that she never finds out. My eyebrows are beautifully thick and curved, incidentally.

A thin brass ring goes bouncing down steps noisily. And still the orchestra is about to agree.

Bandaged, I am something that has been pushed out of Merrett's body in his sleep. Although I can run and jump I have no head at all. I think I am yellow.

What appears to be human hair hangs in long ropes, caked with runs of white sediment, from the gantries the ship must pass under to get to the sky again. Just a few of these ropes of hair, fairly easy to avoid.

There is dangling in the little concrete laboratory, too, from surprised fingers. Twisted black, like monkey limb or hair set in bitumen, stuck to itself and dried, and easy to tear. Blander to taste than anyone would think; its smell of sweet banana spreads everywhere.

Flakes of skin on the gantries, flakes of cellulose adhesive coming off in the evening sun. Flakes of grey paint. That grinning dog that eats everything.

There is hair fringing some of these girders that is soft and fair, and combed out. Eyelash hair that curls in the sunlight and flickers in the wind. Once: a belly rubbed with lemon peel. Somebody's.

Cold in the afternoons, cold in the evenings. There is one eye now, stitched open with wire. Flocks of children fall away when I rub my fingers, and ranks of square houses. How big my fingers are.

What does she think when I rub my tanned fingers down the white doorpost, altering the surface; or across one another? What does she think I am doing? Children with dirty faces, inquisitive mouths hanging open; they are always silent. These fingers I rub are hard, the skin feels dead. What does she think they are?

Trapped somehow, arrested in this doorway, dropped on to one knee, this hand that troubles me resting on the other, thumb and fingers crossing, rubbing. Otherwise peaceful. But arrested here. And I am aware of Amy watching me from very close. She has made herself like a rubber moon. And Dougal; and Joyce behind him. And Merrett squats close to me, looking at me like a friend. They are taller, and everything is narrower, bulkier, softer. As I feel them watching me I become incapable of watching them at all. I wonder what they see when they look at my fingers moving.

The light on the black arms of the big machine shows the edges up hard. The iron arms go down across darkness, into broken light, back through darkness and up again, their edges hard. A sizable piece of worn-looking white cotton stuff is being passed to and fro inside the machine, but it doesn't get dirty.

The face of the stretched undervest. I am not plagued by it, but I know it is there, and its opposite, and its less obvious variants. What is the opposite of a face?

Now there's this trumpet-player, Henrik, come out of the sickbay at last. He looks at me when I'm down, too, friendly, though I never visited him. Cadaverous, sallow man, with cropped grey hair and big moulting brown eyes that weaken his mouth as they move.

I cannot tell how it is lit. In those moments when suddenly I am trapped, or down, and they are looking at me, I am aware of nothing but their compassionate eyes. And when it is gone I cannot imagine how the scene was lit. Not by daylight. Something beyond the doorway and the dark panelling.

Horsehair, the padding of an old chair, pulled out in a flat tangle. Sometimes a man's star stains like a cigarette burn on yellow wood.

Suffering and love in Henrik's eyes. A thinking love. Temptation to make him happy, then outwit him.

What does she think when she knows they are close to me?

There is clear brown glass, there are pink flowers, there is the dark panelling. The distance of late afternoon sun.

A long hand with a tremor rests on the metal arm of a black machine. The thumb rubs drily across the iron, across the side of the forefinger.

A journey. Between Amy's breasts by caterpillar tractor. And back again.

The white porous earth-cloud that passes me through itself, and through; that brings forward more of itself to grow round me, the sounds of factories muffled above my head, beneath my head. To subside through this cloud-bread that puts blind and deaf distance about me.

If she knows they are close to me, she knows she, too, has come close. She sees what Henrik sees when he talks to me.

While Henrik talks to me the others talk to my actor up on deck. With Henrik there is clear brown glass, there are pink flowers, dark panelling. On the land too, there are these things.

A couple of hours before dawn the dry, porous grey fog, webbed with black net. Caked carbon of burnt hair on the lace, sweet smells of the garden at night.

Morning. The actor watching me whenever I go up on deck.

I have been to this convalescent woman's cabin twice already, once in the afternoon, once in the night. She knows I have nothing to do. Her reddish-brown hair is dyed from grey, I think. She's sallow, and rather bony; not exactly elderly, but careful about how she moves, and a little deaf. The first time, the late afternoon sun beat up off the sea through the little curtain; the second time, the faint light from the wall fitting spread out across the panelling without reaching the corners. She watches everything with her brown eyes — doesn't like closing them. In the cabin she says very little. When I'm not with her I don't need to think about her at all. I like a pretty silk frock she has, patterned in grey and red.

The old man on the boat deck no longer says anything as I pass. I go up close and peer at him. Behind the sunglasses his eyes are shut. His hands are folded over his stick, their slight tremors shaking it.

In the swimming pool the actor goes idly along on his back, the sunlit water patterning his breast and belly.

There was a little old man I helped to nurse while he was dying. His paralysed legs grew soft and feminine, his whole body coy. In the coffin he was rouged and decked out in satin frills and ribbons.

The skin of her bared upper arms is pale and flabby, though she is thin. My fingers and my thumbs detect the muscles under the skin. There is a vacancy on her body that her pretty scent does not cover. Across the vacancy her brown eyes follow me.

The water supports the actor, the sun nourishes him, the air delights in his body. The water sprinkles him with stars as he wallows playfully on his back, dipping from side to side.

Suppose she wears nightdresses like the heavy shiny pink one always, not just when I visit. That would be no joke.

The actor's nipples are like the soft oval tops of little puddings flavoured with the palest chocolate, buried in his breast. His flesh absorbs, creamy and riddled with muffled journeys. He is cumulus. The thought of his flesh is a thought rubbed with oranges, painted with honey.

Will she darn the seam we burst open under the arm or will she leave it torn till next time? I know where she keeps her glasses when she takes them off, I can't pretend to myself that I don't. Am I kind to her? Would I be flattered to see the other men she has settled for before me? I think I should not.

The smell of leather passes heavily on the wind to starboard and disappears ahead.

She seems to enjoy me as if she were enjoying something I should not myself like: a shiny, sticky iced cake, for example. That is the newness. If I go many more times she will start to notice me.

Two late afternoons, one after the other. Some repetitions, some variations. Not enough data to set patterns. I play all the time with the simple yes-no toy of whether I go for the second midnight in succession or leave the first to stand as an emergency. Plenty to think about. Nothing gives me a lead.

He flames under the waves in flakes like the setting sun. His navel is an ancient mouth. His teeth strengthen his followers' fingertips as they brush them across his opened lips. His feet, etc. His privy member, handsomely formed, is still as a lizard.

There are circles of beauty across his body. Smiles of beauty across his shoulders. All this has been prepared for: white canvas screens stand about on the deck, flapping tightly in the sun.

Patted dry, he puts on his crisp white jacket again. His fingers, made for smoking cigarettes, settle his collar. He wears his glasses.

Joyce is taller than I thought she was.

Why should they have been doing it in the washroom instead of the sick-bay? Behind the rubber-stoppered door. I had assumed him dead without expecting a funeral. So many other ways. Disposal. But to glimpse the white suit, yellowed, crouched on a stool by the basin while the nurse and the sick-berth attendant were taking the bandages from the head.

The captain has gone down into the depths of the ship; into slippery wet blackness, invisible by day.

I could go to her in the mornings too. Or instead. Perhaps somebody else goes in the mornings. But I was the first, yesterday. She'd had no time after being so ill to get anybody else.

Bent over on to the porcelain it looked at me without any change of expression. Placid, interested in what was happening. Discernibly a human head, bald, with one big eye looking at me. Where the dressing peeled away I could see the contusion, the spontaneous breaks in the skin, the wounds that looked too tired to bleed. All the time he concentrates on his head.

For a moment, a great shaking glass sheet for a window, with the blurred pink figure of a man shouting at me through it. But I cannot hear a sound from him or from the wind that distorts him so.

Dougal comes past. I tell him to avoid the washroom. He thanks me.

She wants to make me forget. That is what she says, starting to notice me. In the darkness, slippery nylon wrinkled against my face, my head full of cold scent, I feel self-pity coming on.

What is it she thinks I can remember? I can never remember enough. Against her side I rub the chain I feel growing in me, but she misunderstands. I want to be kind to her.

Above us, below us, the ship spins slowly in the night, grinding quietly. From a porthole, Merrett surveys the waters with distrust.

White plastic fibrils appear here and there in the darkness of the ship. They stretch and snap and bud, then break quickly and disappear.

The water comes over, brimming, golden-grey, with nobody to notice it. Full of pale light towards the surface. No more taste to it than to a human body. Broken thistles afloat in it. Lapping softly down the plates of the hull it finds its level again.

In the dawn light, something dark hopping from one wave trough to the next, keeping level with the ship, just as a sunbeam sometimes does. Nothing to bother about.

The ship draws in again the coils of piping that trailed beneath its hull in the night. Far from land, we sail in shallows where grey cylinders and globes lie under the water, near enough to the surface for their rivets to show.

I go below, and feel the ship above me turn, turning, trying to find the night again.

This is what it is like on the land: the town-gods, with coloured rings painted round their eyes, drive their cars down to the water's edge and stand in them watching the ships go by.

Protection has its grip slowly peeled off all of us, and off all our things. We're left glistening and tight. What it is that will taunt, what is it that will lunge?

On the land there are big old sheds of corrugated iron, their reddish-brown paint much faded. Cinder paths behind them, and tough grey-green weeds.

The white iron ceiling, the ivory corner, the washroom door. They have finished in there. The room has been full of steam; the mirrors and walls are clouded over. Damp soiled dressings in the bin. They have gone away. I wonder whether they have forgiven him.

With steam in my glass, a wet clock with fingers that keep slipping back, the effort of propping my shoes against a slippery tiled wall, things are unsteady. Iron ladders lead down, you can see feet, waists, heads moving about. Wiping the moisture off a chromium rim of this chromium-rimmed thing gives a narrow strip of mirror; and glinting, scissor-eyed, they look in from it at what's going on. Two or three of them, not recognizable. Here there are many black shoes, shuffling, swivelling, some of them women's. Wearing at the linoleum.

I smell methylated spirits. Dougal there, just behind me, lying as if dead. He's stripped to the waist but his trousers are neat. Amy, kneeling beside him, pours the stuff on to wadding and wipes his chest and throat. People tread on them in passing. Ducts up into the air, down into the dark. More feet, blackshod, bare pink, thudding close to me along the metal wall beside my ear or across the thick reeded glass lid through which I look up into the bright fog of daylight. High up in it, billows of orange smoke seem to be going past.

OXFORD POETS

Fleur Adcock

James Berry

Edward Kamau Brathwaite

Joseph Brodsky

D. J. Enright

Roy Fisher

David Gascoyne

David Harsent

Anthony Hecht

Zbigniew Herbert

Thomas Kinsella

Brad Leithauser

Herbert Lomas

Derek Mahon

Medbh McGuckian

James Merrill

John Montague

Peter Porter

Craig Raine

Tom Rawling

Christopher Reid

Stephen Romer

Carole Satyamurti

Peter Scupham

Penelope Shuttle

Louis Simpson

Anne Stevenson

George Szirtes

Anthony Thwaite

Charles Tomlinson

Chris Wallace-Crabbe

Hugo Williams

also

Basil Bunting

W. H. Davies

Keith Douglas

Ivor Gurney

Edward Thomas